Make:
Getting Started with Raspberry Pi

4TH EDITION

Getting to Know the Inexpensive ARM-Powered Linux Computer

Matt Richardson, Shawn Wallace,
and Wolfram Donat

Make:

Getting Started with Raspberry Pi, 4th Edition
by Matt Richardson, Shawn Wallace, and Wolfram Donat

Published by Make Community, LLC
150 Todd Road, Suite 100, Santa Rosa, CA 95407

Make: books may be purchased for educational, business, or sales promotional use. Online editions are also available for most titles.
For more information, contact our corporate/institutional sales department: 800-998-9938

Publisher: Dale Dougherty
Editor: Patrick DiJusto, Michelle Lowman
Copy Editor: Craig Couden
Interior Designer: David Futato
Cover Designer: Juliann Brown
Illustrator: Rebecca Demarest

December 2012: First Edition
October 2014: Second Edition
July 2016: Third Edition
September 2021: Fourth Edition
Revision History for the Fourth Edition: 10/29/2021

See www.oreilly.com/catalog/errata.csp?isbn=9781680456998 for release details.

978-1-680-45699-8

O'Reilly Online Learning

For more than 40 years, www.oreilly.com has provided technology and business training, knowledge, and insight to help companies succeed.

Our unique network of experts and innovators share their knowledge and expertise through books, articles, conferences, and our online learning platform. O'Reilly's online learning platform gives you on-demand access to live training courses, in-depth learning paths, interactive coding environments, and a vast collection of text and video from O'Reilly and 200+ other publishers. For more information, please visit www.oreilly.com

How to Contact Us:

Please address comments and questions concerning this book to the publisher:

Make: Community

150 Todd Road, Suite 100, Santa Rosa, CA 95407

Make: Community is a growing, global association of makers who are shaping the future of education and democratizing innovation. Through *Make:* magazine, and 200+ annual Maker Faires, *Make:* books, and more, we share the know-how of makers and promote the practice of making in schools, libraries and homes.

You can send comments and questions to us by email at books@make.co.

To learn more about *Make:* visit us at make.co.

Contents

What They're Saying

...about *Getting Started with Raspberry Pi, 4th Edition*

"An exceptional introduction to the Raspberry Pi, accessible to a beginner and with in-depth references for experienced makers." —*Tim Wright, Aerospace Engineer*

"I'm a kid from New York City. I thought the book was really helpful with the command line. I wanted to do some stuff with a project, but I was quite intimidated. Now that I read this book, I can set off!" —*Kenji D., age 11.*

Preface

Ten years.

It's been ten years since the Raspberry Pi was first announced in 2011. And what a decade it has been. A credit-card–sized computer for $35? That I can hook up to my existing monitor and keyboard setup? And connect to physical things via GPIO pins? It seemed like a pipe dream. This is why, when it started shipping, the Raspberry Pi created a frenzy of excitement.

Demand outstripped supply for months, and the waitlists for these mini computers were very long. Some of their newest products (and not-so-new) like the Pi Zero W and the Pi 4 still have limited availability; it's difficult to find a place that will sell more than two Pi Zeros to a customer. Besides the price, what is it about the Raspberry Pi that tests the patience of this hardware-hungry mass of people? Before we get into everything that makes the Raspberry Pi so great, let's talk about its intended audience.

Eben Upton and his colleagues at the University of Cambridge noticed that students applying to study computer science didn't have the skills that they did in the 1990s. Students were considering themselves skilled at what they called "computer science" when all they could do was use MS Word and Excel and perhaps write a little HTML and perhaps JavaScript.

Upton and the others attributed this to—among other factors— the "rise of the home PC and games console to replace the Amigas,

BBC Micros, Spectrum ZX, and Commodore 64 machines that people of an earlier generation learned to program on."[1]

Because the computer has become important for every member of the household, it may also discourage younger members from tinkering around and possibly putting such a critical tool out of commission for the family. Parents don't want their children "hacking" the family computer while learning to program, because theyrun the risk of possibly breaking it.

Meanwhile, mobile phone and tablet processors had become less expensive while getting more powerful, clearing the path for the Raspberry Pi's leap into the world of ultra-cheap-yet-serviceable computer boards. The ARM chip family that's used in all of the Pi boards got its main start inside mobile phones.

As Linus Torvalds, the founder of Linux, said in an interview with BBC News, the Raspberry Pi makes it possible to "afford failure." If a child (or an adult) manages to brick the Pi, he or she can just buy another one, for far less than the cost of replacing a laptop.[2]

Raspberry Pi Foundation

It's important to note that Raspberry Pi primarily exists to advance the charitable mission of the Raspberry Pi Foundation. That mission is to "put the power of computing and digital making into the hands of people all over the world." The Raspberry Pi Foundation hopes that people—kids especially—will learn to code, learn how computers work, and learn how to make things with computers.

With every Raspberry Pi purchase you make, you're not only paying for the cost of the hardware, fulfillment, and the engineering behind it, you're also contributing the free online resources, free teacher training, and special programs that the Raspberry Pi Foundation offers to further its charitable mission.

As you'll learn in this book, the Raspberry Pi is great for learning, but it also makes for a powerful tool. Even if the primary

---

1 "About us," Raspberry Pi Foundation (www.raspberrypi.org/about).
2 Leo Kelion, "Linus Torvalds: Linux Succeeded Thanks to Selfishness and Trust," BBC News, June 12, 2012.

purpose of the board is for education, we find that its utilization stretches into commercial and industrial applications. Companies use it for things such as sensor networks, remote monitoring, and product prototyping. Even though the Raspberry Pi is great for kids, you should keep in mind that it's a real computer. It's not a toy or some kind of watered-down device.

What Can You Do with It?

One of the great things about the Raspberry Pi is that there's no single way to use it. Whether you just want to watch videos and browse the Web, or you want to hack, learn, and make with the board, the Raspberry Pi is a flexible platform for fun, utility, and ex-perimentation. Here are just a few of the different ways you can use a Raspberry Pi:

General-purpose computing

It's important to remember that the Raspberry Pi is a general-purpose computer and you can, in fact, use it as one. The Pi version 4.0 (the most recent release as of this writing), with 8GB of RAM and two HDMI outputs capable of 4K 60fps, is power-ful enough to completely replace most general users' desktop computers. After you get it up and running in Chapter 1, you can launch a web browser to access email, news sites, and social networks, which is a lot of what we use computers for these days. Going beyond the Web, you can launch the free and open-source LibreOffice (www.libreoffice.org) productivity suite, which allows you to work with documents and spreadsheets when you don't have an internet connection.

Learning to program

Because the Raspberry Pi is meant as an educational tool to encourage kids to experiment with computers, it comes pre-loaded with interpreters and compilers for many different pro-gramming languages. If you're eager to jump into writing code, the Python programming language is a great way to get started, and we cover the basics of it in Chapter 4. But with Raspberry

Pi, you're not limited to only Python. You can write programs for your Raspberry Pi in many different programming languages, including C, Ruby, Java, and Perl.

All newer versions of the Raspberry Pi OS come pre-installed with Scratch, a programming environment meant to introduce younger users to programming concepts. There's even a programming language and development environment for creating music called Sonic Pi.

Project platform

The Raspberry Pi differentiates itself from a regular computer not only because of its price and size, but also because of its ability to integrate with electronics projects. Starting in Chapter 6, we'll show you how to use the Raspberry Pi to control components from LEDs to AC devices, and you'll learn how to read the state of buttons and switches.

Product prototyping

More and more electronics products use Linux computers inside, and now this world of *embedded Linux* is more accessible than ever. Let's say you create something with your Raspberry Pi that would make a great product for the every-day consumer. With the *Raspberry Pi Compute Module* (a smaller version of the board that we'll discuss later), it becomes possible to create a product that's powered by Raspberry Pi. Companies can also use the smaller Pi versions like the Zero and Zero W for products and prototypes where a full-size Pi is either too big or too expensive to make a good fit.

Raspberry Pi for Makers

As makers, we have a lot of choices when it comes to platforms on which to build technology-based projects. Microcontroller development boards like the Arduino (and the new Raspberry Pi Pico RP2040) have long been a popular choice because they've become very easy to work with. But *system on a chip* platforms like the Raspberry Pi are a lot different than traditional microcontrollers in many ways. It's a common misconception that the Pi and a microcontroller are interchangeable; in fact, they are completely different devices that fulfill totally different functions. The Raspberry Pi is a computer, like your desktop computer, while the Arduino is a microcontroller that has no business trying to replace a Dell or an iMac.

This is not to say that a Raspberry Pi is *better* than a traditional microcontroller; it's just different. For instance, if you want to make a basic thermostat, you're probably better off using an Arduino Uno or similar microcontroller for purposes of simplicity. But if you want to be able to remotely access the thermostat via the Web to change its settings and download temperature log files, you should consider using the Raspberry Pi.

Choosing between one or the other will depend on your project's requirements, and, in fact, you don't necessarily have to choose between the two. In Chapter 5, we'll show you how to use the Raspberry Pi to program the Arduino and get them communicating with each other. Many projects—built by both hobbyists and professional engineers—require both a controlling CPU like the Pi and a microcontroller like the Arduino.

As you read this book, you'll gain a better understanding of the strengths of the Raspberry Pi and how it can become another useful tool in the maker's toolbox.

But Wait... There's More!

There's so much you can do with the Raspberry Pi, it couldn't fit into scores of books, much less just one. Here, for example, is a list of several other things that can be done quite easily with the Pi:

Media center

Because the Raspberry Pi has HDMI outputs (and composite video embedded in the 3.5mm AV jack), it's easy to connect the Pi to almost any modern television. As I mentioned before, it's also quite capable of displaying full screen video in 1080p or even 4K resolution. So it seems natural that you may want to leverage these capabilities and make the Pi a media server. You can run the Plex media service on it, and you can install media player operating systems like Open ELEC (openelec.tv) and OSMC (osmc.tv) on the Pi. These systems can play lots of different media formats and are designed to be easily accessible on a large TV.

"Bare-metal" computer hacking

Most programmers write code that runs within an OS like Windows, Mac OS, or Linux. However, it's possible to write code that runs directly on the processor, similar to a microcontroller, or within a different type of operating system called a *real-time operating system* like FreeRTOS. Although it's not for a beginning user, you can write and run these programs on the Pi, or even write your own operating system! A free online course from the University of Cambridge (bit. ly/1BW2e3C) walks you through the process of developing an OS, though it's a bit outdated now and has not been updated for newer versions of the Pi.

Retro gaming

If you're a retro gaming enthusiast who misses the days of Super Mario Bros. and Joust and Galaga and others, you can use the RetroPie (retropie.org.uk) as a platform for emulating a lot of older gaming systems like the Nintendo and the Atari. You can use after-market add-ons for the Pi (called HATs) to wire up joysticks,

Linux and Raspberry Pi

Most general computers that you're used to using run an operating system like Windows, MacOS, or Linux. The operating system is what you use to interact with the applications and programs that run on the computer, and it could be said that the OS acts as a "buffer" between the user or programmer and the hardware—you don't need to know the specifics of the TCP/IP protocol or the Ethernet chip to program and use a web browser, for instance.

The Pi is no different and runs a flavor of Linux called the Raspberry Pi OS. It's a Debian-based distribution, and if you've ever used Debian or Ubuntu in the past, the Pi's OS will seem very familiar to you. It's a great match for the Pi because it's free, it's easy to use, and it's also hackable.

But you aren't limited to the Pi OS. There are other distributions you can load onto the Pi, such as Ubuntu, and even some non-Linux systems like Windows 10 Core or even Android. Check out Chapter 3 for a rundown of some of the options available to you. In this book, I use the standard Raspberry Pi OS available from the Pi Foundation's download page (www.raspberrypi.org/software). It's a good place to start. And if you're not familiar with Linux, check out Chapter 2 for a quick introduction to this surprisingly easy-to-use OS.

What Others Have Done with Raspberry Pi

If you've gotten yourself a Pi and are having trouble deciding what to do with it, fear not. There are so many projects out there to build with your Pi it'd be silly for me to try and choose just a few to share with you. Seriously—if you can think of something to do with the Pi, there's an excellent chance someone else has done it already, whether it's a weather station or a planetary rover or an arcade game or even a supercomputing cluster. But don't let the fact that you're not the first discourage you; rather, be happy that the community that has sprung up around the Pi and its projects is immense and supportive and no matter what you do with your Pi you'll be contributing and adding to an incredibly diverse ecosystem.

Conventions Used in This Book

The following typographical conventions are used in this book:

Italic

Indicates new terms, filenames, and file extensions.

`Constant width`

Used for program listings, as well as within paragraphs to refer to program elements such as variable or function names, databases, data types, environment variables, statements, and keywords.

`Constant width bold`

Shows commands or other text that should be typed literally by the user.

`Constant width italic`

Shows text that should be replaced with user-supplied values or by values determined by context.

 This element signifies a tip or suggestion.

 This element signifies a general note.

 This element indicates a warning or caution.

Acknowledgments

I'd like to thank a few people who have contributed to this edition of *Getting Started with Raspberry Pi*:

First and foremost, the original authors—Matt Richardson and Shawn Wallace. Their original book was (and still is) an awesome introduction to this little computer and I'm honored to be able to add my thoughts and work to it.

And of course, Patrick Di Justo and the rest of the new Make: team for thinking of me when it came time to update the book again.

Bill of Materials

Following is a list of the major components used in this book:

- Raspberry Pi (obviously)
- Raspberry Pi Pico
- microSD card (at least 8GB)
- Power supply (3A or greater, preferably)
- HDMI cable(s)
- HDMI-micro
- HDMI adapter
- mouse
- keyboard
- webcam and/or Pi camera module
- powered USB hub, either 2.0 or 3.0
- Pi case
- Arduino (doesn't matter which one)
- solderless bread board
- assortment of jumper wires (M/M, M/F, and F/F)
- assorted LEDs
- push button switch
- assorted resistors
- Power Switch Tail
- IIADS1115 or ADS1015 ADC board
- potentiometer
- force-sensitive resistor
- photocell (light-sensitive resistor)

1/Getting Up and Running

A few words arise over and over when people talk about the Raspberry Pi: small, cheap, hackable, education-oriented. However, it would be a mistake to describe it as *plug-and-play*, even though it is easy enough to plug the Pi into a TV set and get something to appear on the screen. The bottom line is that the Raspberry Pi is not a consumer device, and depending on what you intend to do with it, you'll need to make a number of decisions about peripherals and software when getting up and running.

Of course, the first step is to actually acquire a Raspberry Pi. Chances are you have one by now, but if not, the Raspberry Pi is available from a huge selection of vendors online, including Maker Shed (makershed.com), Sparkfun (sparkfun.com), Adafruit (adafruit. com), Element14 (www.element14.com/community/welcome), RS Components (www.rs-online.com), DigiKey (digikey.com), and even good old Amazon.

The Raspberry Pi's low price is obviously an important part of its story. Enabling the general public to go directly to a distributor and order small quantities of a computer for the same price offered to resellers is an unusual arrangement. A lot of potential resellers were confused by the original announcements of the price point; it was hard to see how there could be any profit margin. That's why you'll see some resellers (particularly on Amazon and eBay) adding a slight markup to the price of any of the Pi models. Some of these resellers, such as Adafruit and Spark-fun, offer a whole host of accessories to go with the Pi, including HATs (addon boards to immediately enable certain functionality for the Pi), LCD- and touchscreens, and various cases. While these resellers may mark up the price a bit, it's often worth it (in my opinion) to pay a little more and browse the add-ons that they have made available to the average hobbyist.

Enough micro economic gossip; let's start by taking a closer look at the Raspberry Pi board.

A Tour of the Boards

There have been quite a few different versions of the Raspberry Pi board. The first version was the Raspberry Pi 1 Model B, which was followed by a simpler and cheaper Model A. In 2014, the Raspberry Pi Foundation announced a significant revision (and improvement) in the board design: the Raspberry Pi 1 Model B+. The Model B+ set the form-factor for "mainline" Raspberry Pis for the foreseeable future. Since then, the Foundation has also created a device for embedding the Pi in products, called the Compute Module. In 2015, it also released a stripped-down $5 model called Raspberry Pi Zero, followed closely by a wireless version called the Raspberry Pi Zero W. In February 2016, the Raspberry Pi 3 Model B came online, then the Model 3B+. As of this writing, the latest board is the Raspberry Pi 4B, which is a huge step up in terms of the Pi's capabilities. There are three different models of the 4 available, depending on how much RAM you want to play with: the 2GB model costs $35 (the same price as the original Pi), the 4GB costs $55, and the 8GB—the newest model—costs $75. Although that may seem a little high for a Pi, when you consider that it's powerful enough to conceivably replace your usual desktop computer, $75 seems like a steal.

Over the years, there have been a few different versions of the mainline Raspberry Pi, which is the $35 model with four USB ports that most people tend to use. Each of these models added performance improvements to the processor. Raspberry Pi 2 added more RAM, the Raspberry Pi 3 added onboard Wi-Fi and Bluetooth, and the Pi 4 added even more RAM and more display capabilities, along with USB 3.0.

If you're following along with the examples in this book, any of these mainline Raspberry Pis will do just fine.

Mike Senese

Figure1-1. *Raspberry Pi 2, 3, and 4 (Model B), from top left*

Let's start with a tour of what you'll see when you take your Raspberry Pi out of the box.

It's easy to think of Raspberry Pi as a microcontroller development board like Arduino or as a laptop replacement. In fact, the Pi is more like the exposed innards of a mobile device with maker-friendly headers for various ports and functions. Figure 1-2 shows the parts of the board.

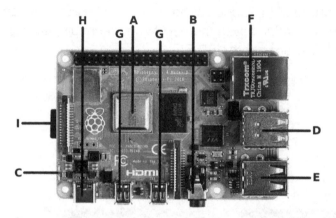

Figure 1-2. *A map of the hardware interface of the Raspberry Pi*

Here's a description of each part:

A. *The processor*
   At the heart of the Raspberry Pi is the same kind of processor you'd find in a cell phone. If you're using Raspberry Pi 4, this is a 64-bit, quad-core 1.5GHz system on a chip, which is built on the ARM A72 architecture. ARM chips come in a variety of architectures with different cores configured to provide different capabilities at different price points. Raspberry Pi 1 had 512 megabytes of RAM and Raspberry Pi 2 and 3 have one gigabyte of RAM. The Pi 4, as I mentioned earlier, comes in flavors of 2, 4, or 8GB of RAM.

B. *Composite video and analog audio out*
   Analog audio and video outputs are available on a standard 3.5mm 4-pole plug connector. You can pick up A/V to RCA conversion cables that will fit the 3.5mm jack on your Pi from many of the vendors listed on page 1.

C. *Status LEDs*

Two indicator LEDs on the board provide visual feedback (Table 1-1). There are also network activity LEDs on the Ethernet port itself.

Table1-1.*The status LEDs*

ACT	Green	Lights when the SD card is accessed
PWR	Red	Hooked up to 3.3V power

 Starting with Raspberry Pi 3, the status LEDs are placed near the MicroUSB power port as shown in Figure 1-2. For previous boards, you'll find them near the GPIO pins.

D. *External USB 3.0 ports*

New to the Pi 4 are two USB 3.0 ports, differentiated by the blue color of the ports (a shade called Pantone 300C, if you're interested). USB 3.0 is not only capable of providing more electrical power to external peripherals (provided you are sourcing more power to the Pi itself), it also offers up to 10x faster data transfer speeds than USB 2.0.

E. *External USB 2.0 ports*

On all versions of the Raspberry Pi, there are at least two USB 2.0 ports for connecting peripherals like keyboards, mice, thumb drives, and printers. While many USB devices can be powered from these ports, you may want to consider using a powered external hub if you have peripherals that need more power, such as a hard drive.

F. *Ethernet port*

This is a standard RJ45 Ethernet port capable of 1 gigabit per second data speed. Connect this to your router to get online; otherwise, you can use Wi-Fi—the onboard dual-band Wi-Fi chip is compatible with b, g, n, and ac Wi-Fi bands.

G. *Micro HDMI connectors*

The two micro HDMI ports on the Pi 4 are each capable of 60fps 4K video output. (Video performance may vary on how much you're taxing the Pi; just because the GPU is capable of these outputs doesn't mean it will always look good.)

H. *Power input*

There is no power switch on the Pi. This USB-C connector is used to supply power (and only power; this isn't an additional USB port). USB-C was selected because the connector is cheap and USB power supplies are easy to find. When you're powering your Pi 4, the Pi Foundation strongly suggests a power supply capable of delivering 3A (15W), particularly if you're planning on powering external devices as well.

I. *The microSD card slot*

There's no hard drive on the Pi. Everything from the operating system to working programs to data are stored on a microSD card. Raspberry Pi 1 and 2 are equipped with spring-loaded full-size SD slots, so you'll push to put the SD card in and push again to take it out. With Raspberry Pi 3, they did away with the spring-loaded component in favor of a friction-fit microSD slot. On that model and all those following, including the Zero and the Zero W, you'll push to insert the microSD card and pull to remove it. Of course, you should only insert or remove the SD card when the Raspberry Pi is powered down.

Figure 1-3 shows all of the power and input/output (I/O) pins on the Raspberry Pi.

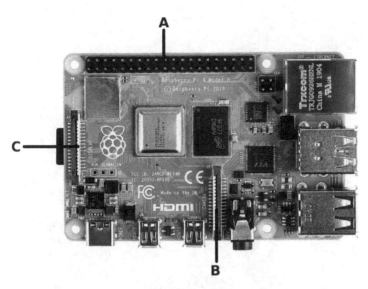

Figure 1-3. *The pins and headers on the Raspberry Pi*

Here's a description of the pins and headers shown:

A. *General-purpose input/output (GPIO) and other pins*
The current Raspberry Pis have a 2×20 pin GPIO header. Chapters 6 and 7 show how to use these pins to read buttons and switches and control actuators like LEDs, relays, or motors.

B. *The Camera Serial Interface (CSI) connector*
This port allows a camera module to be connected directly to the board (see Figure 1-4).

C. *The Display Serial Interface (DSI) connector*
This connector accepts a 15-pin, flat ribbon cable that can be used to communicate with the official Raspberry Pi touch display.

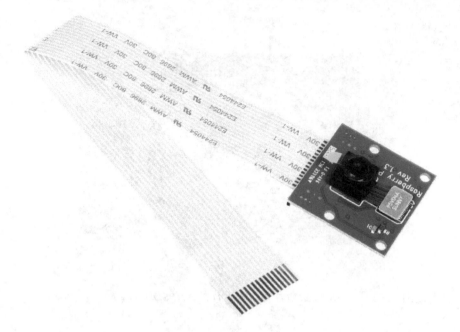

Figure 1-4. *The Raspberry Pi camera module connects directly to the CSI connector. See Chapter 9 for a full discussion of cameras and the Pi.*

The Proper Peripherals

Now that you know where every thing is on the board, you'll need to know a few things about the proper peripherals to use with the Pi. There are a bunch of prepackaged starter kits that have well-vetted parts lists, though there are also a few caveats and gotchas when fitting out your Raspberry Pi. There's an extensive list of supported peripherals (elinux.org/RPi_VerifiedPeripherals) on the eLinux.org wiki, but these are the most basic:

A power supply

This is the most important peripheral to get right. For the Raspberry Pi 4, you should use a USB-C adapter that can provide 5V and at least 3,000mA (3A) of current. If you're using a Pi 3/B+, you'll need a micro-USB adapter that can supply 5V and at least 1.5A of current. Older Pis will only need 1,000mA (1A). A cell phone charger won't necessarily cut it, even if it has the correct connector. Many cellphone chargers don't provide enough current, so check the rating marked on the back. An underpowered Pi may still seem to work but will be flaky and may fail unpredictably. Newer versions of the Pi will also blink an error message on the desktop — "Under voltage detected!" If in doubt, use the official Raspberry Pi power supply, which is available at most places where Raspberry Pis are sold.

 There are also several battery-pack solutions for taking your Raspberry Pi on the go; the same rules about voltage and current apply there as well.

With the current version of the board, it is possible to power the Pi from a USB hub that feeds power back into one of the two external USB ports. However, there isn't much protection circuitry, so it may not be the best idea to power it over the external USB ports. This is especially true if you're going to be doing electronics prototyping where you may accidentally create short circuits that may draw a lot of current.

A microSD card
You'll need at least 8GB, and it should be a Class 10 card for the best read and write performance. There are operating systems that fit onto SD cards with less than 8GB, but the standard Raspberry Pi OS installation requires at least an 8GB microSD card.

USB keyboard and mouse
They'll be helpful for controlling your computer. These peripherals are fairly generic, so no need to use anything fancy.

HDMI cable(s)
If you're connecting to a monitor, you'll need this or an appropriate adapter for a DVI monitor. If you're using the Pi 4, you'll need either an HDMI/HDMI microcable or the appropriate adapter. You can also run the Pi headless, as described later in this chapter. HDMI cables can vary wildly in price. If you're just running a cable three to six feet to a monitor, there's no need to spend more than $3 on an HDMI cable. If you are running long lengths

of cable or displaying 4K video, you should definitely research the higher-quality cables and avoid the cheap generics.

Ethernet cable
Your home may not have as many wired Ethernet jacks as it did five years ago. Because everything is wireless these days, you might find the wired port to be a bit of a hurdle; see the section "Getting Online" on page 22 for some alternatives to plugging the Ethernet directly into the wall or a hub.

Wi-Fi USB dongle
If you're using one of the older Pis, you may want to add a Wi-Fi dongle for wireless internet access. Many 802.11 WiFi USB dongles work with the Pi out of the box. Wi-Fi uses a lot of power, so you will need to make sure you have anadequate power supply; a 2A supply or a powered USB hub is a good choice. If you are having problems with a Wi-Fi dongle, power is almost always the problem.

You may also want to consider some of the following add-ons:

A powered USB hub
If you want to add more than four USB devices to a mainline Raspberry Pi, you'll need a USB hub. A powered USB 2.0 hub (or 3.0 for the Pi 4) is recommended.

Heatsink
A heatsink is a small piece of metal, usually with fins, that creates a lot of surface area to dissipate heat efficiently. Heat sinks can be attached to chips that get hot. The Pi's chipset was designed for mobile applications, so a heat sink isn't necessary most of the time. However, as we'll see later, there are cases where you may want to run the Pi at higher speeds, or crunch numbers over an extended period, and the chip may heat up a bit. Some people have reported that the network chip can get warm as well.

Real-time clock

You may want to add a real-time clock chip (like the DS1307) for logging or keeping time when offline. This is also necessary should you want to experiment with running a real-time operating system on the Pi.

Camera module

A $25 Raspberry Pi camera module is available as an official peripheral. Version 2 of the camera sports an 8MP image sensor and is capable of recording 1080p video, while the newest version has a 12MP sensor. There is also an infrared camera model; with a few infrared LEDs to "light" the scene, you can take pictures in total darkness. You can also use a USB webcam (more on this in Chapter 9).

LCD

Most Liquid Crystal Displays can be used via a few connections on the GPIO header. Look for a TFT (thin-film transistor) display that can communicate with the Pi using the SPI (Serial Peripheral Interface) pins on the header. The Raspberry Pi Foundation also has a touch display that connects to the DSI Interface on the Raspberry Pi.

Soundcards

You'll probably find that the built-in analog audio is inadequate for most of your projects. If you want high-quality sound output (or input) from the Pi, you'll need a soundcard. Many USB soundcards also work well with the Pi; Behringer's U-Control devices are a popular, inexpensive option.

Laptop dock

Several people have modified laptop docks intended for cellphones (like the Atrix lapdock) to work as a display/base for the Raspberry Pi. Some companies like Pi-Top create a laptop-like device specifically for Raspberry Pi. (As of this writing, the Raspberry Pi 400 is also available, which is a Pi in a laptop form factor. This design definitely calls for a laptop dock, in my opinion.)

HATs

A number of vendors and open hardware folks have released add-on daughterboards that sit on top of the Pi and connect via the GPIO header. These boards add capabilities like driving LCDs, motors, or analog sensor inputs. If you're familiar with Arduino terminology, you might call these daughterboards "shields," but the Raspberry Pi Foundation calls them HATs (Hardware Attached on Top), see Figure 1-5. The full specification is available on the Raspberry Pi Foundation's GitHub page (github.com/raspberrypi/hats).

Figure 1-5. *The Sense HAT add-on board includes an LED matrix, a suite of sensors, and a joystick input. It was designed for the Raspberry Pis that were sent to the International Space Station.*

To share just one example, the Raspberry Pi Foundation makes a HAT called the Sense HAT, which includes an RGB LED matrix; sensors for temperature, pressure, and humidity; an accelerometer; a gyroscope; and a magnetometer. It also has a five-position joystick. It's the HAT that was designed for the Raspberry Pis that were sent to the Interntional Space Station as part of the Foundation's AstroPi program.

The Case

You may find that you want a case for your Raspberry Pi. The stiff cables on all sides make it hard to keep the Pi flat, and some of the components like the SD card slot can be mechanically damaged even through normal use.

There are a bunch of premade cases available, but there are also a lot of case designs available to download and fabricate on a laser cutter or 3D printer. In general, avoid tabbed cases where brittle acrylic is used at right angles. The layered acrylic of the Pi bow (shop.pimoroni. com/?q=pibow) is a colorful option (Figure 1-6).

The Raspberry Pi Foundation also creates an official case, which uses a nice injection-molded design. It has multiple parts that can be removed to allow access to the GPIO pins and other components. (Figure 1-7).

It should probably go without saying, but it's one of those obvious mistakes you can make sometimes: make sure you don't put your Raspberry Pi on a conductive surface. Flip over the board and look at the bottom; there are a lot of components there and a lot of solder joints that can be easily shorted. Another reason why it's important to case your Pi!

Figure 1-6. *The colorful Pibow case from Pimoroni*

Figure1-7. *With the official Raspberry Pi case, you can remove the top and sides to access the different parts of the board.*

Choose Your Distribution

The Raspberry Pi runs Linux for an operating system. Linux is technically just the kernel, but an operating system is much more than that—it's the total collection of drivers, services, and applications that makes the OS. A variety of flavors or distributions of the Linux OS have evolved over the years. Some of the most common on desktop computers are Ubuntu, Debian, Fedora, and Arch. Each has its own communities of users and is tuned for particular applications.

Because the Pi is based on a mobile device chipset, it has different software requirements than a desktop computer. The Broadcom processor has some proprietary features that require special "binary blob" device drivers and code that won't be included in any standard Linux distribution. And, while most desktop computers have gigabytes of RAM and hundreds of gigabytes of storage, the Pi is more limited in both regards (though the 8GB of RAM available on the Pi 4 rival that available on some lower-end desktop computers). Special Linux distributions that target the Pi have been developed.

In this book, we will concentrate on the official Raspberry Pi OS distribution, which is based on Debian. Note that though raspbian. org still exists, it does not seem to be affiliated with the Raspberry Pi OS, and is a community site, not operated by the Foundation. If you're looking for the official distribution, visit the Raspberry Pi Foundation's downloads page (raspberrypi.org). Other specialized distributions are explored in Chapter 3.

Flash the SD Card

Many vendors sell SD cards with the operating system preinstalled; for some people, this may be the best way to get started. Even if it isn't the latest release, you can easily upgrade once you get the Pi booted up and on the internet.

The easiest way to get the OS on the microSD card is to use the NOOBS tool. Don't take offense; no one is questioning your computer acumen. NOOBS stands for New Out Of the Box Software and is a configuration tool that will help install the OS.

You'll need an SD card (at least 8GB) and reader, then follow these steps: when you boot up the Pi, you'll see a configuration screen with several OS options. Select Raspberry Pi OS and hit the Install button; that's all there is to it!

For Advanced Users: Create Your Own Disk Image

The first thing you'll need to do is download one of the distributions from the Raspberry Pi Foundation's downloads page (www.raspberrypi.org/downloads) or one of the sites in Chapter 3. Note that you can't just drag the disk image onto the SD card; you'll need to make a bit-for-bit copy of the image. You'll need a card writer and a disk image utility; any inexpensive card writer will do. The instructions vary depending on the OS you're running. Unzip the image file (you should end up with a *.img* file), then follow the appropriate directions described in Appendix A.

Faster Downloads with BitTorrent

You'll see a note on the download site about downloading a torrent file for the most efficient way of downloading Raspberry Pi OS. Torrents are a decentralized way of distributing files; they can be much faster because you'll be pulling bits of the download from many other torrent clients rather than a single central server. You'll need a BitTorrent client if you choose this route.

Some popular BitTorrent clients are:

Vuze (www.vuze.com)
Integrated torrent search and download

Miro (www.getmiro.com)
Open source music and video player that also handles torrents

MLDonkey (mldonkey.sourceforge.net)
Windows and Linux-only file sharing tool

Transmission (www.transmissionbt.com)
Lightweight Mac and Linux-only client; also used in embedded systems

Booting Up

Follow these steps to boot up your Raspberry Pi for the first time:

1. Push the microSD card into the socket on the bottom of the board. On Raspberry Pi 1 and 2, it'll click into place. On Raspberry Pi 3 and all later versions, the micro SD card won't click and is held in place by friction.
2. Plug in a USB keyboard and mouse.

3. Plug the HDMI output into your TV or monitor. Make sure your monitor is on and set to the correct input. If you're working with the Pi 4, you'll need a micro-HDMI to HDMI adapter as well.

4. Last, plug in the power supply. It's a good habit to make sure everything else is hooked up *before* connecting the power.

If all goes well, you should see a bunch of startup log entries appearing on your screen. At the top, you'll see a Raspberry Pi logo, or four if you're using a quad-core model (Raspberry Pi 2 or later). If things don't go well, skip ahead to "Troubleshooting" on page 26. These log messages show all of the processes that are launching as you boot up the Pi. You'll see the network interface be initialized, and you'll see all of your USB peripherals being recognized and logged. You can see these log messages after you login by typing `dmesg` on the command line.

Configuring Your Pi

The very first time you boot up, you'll be presented with the Raspbian desktop environment. The first thing you'll want to do is set a few settings with the Raspberry Pi Configuration tool. To open it, click Menu → Preferences → Raspberry Pi Configuration (see Figure 1-8).

Next, in Figure 1-9, we'll show you which configuration options are essential and which you might want to come back to if you need them.

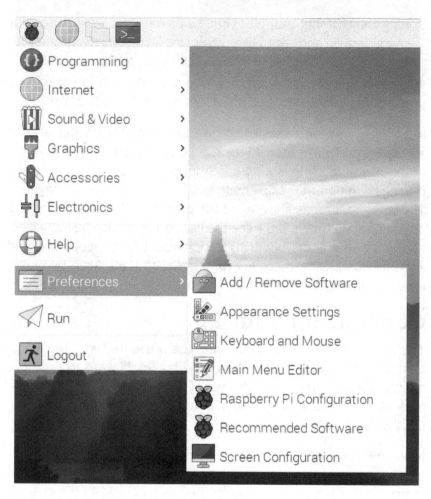

Figure 1-8. *How to launch the Raspberry Pi Configuration tool.*

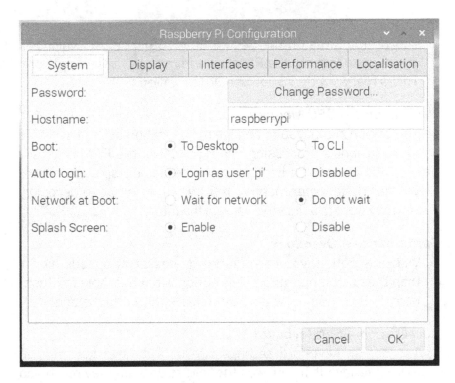

Figure 1-9. *The Raspberry Pi Configuration tool allows you to change many of the important settings on your Raspberry Pi.*

System → Change Password

If you're on a network with others, it's a good idea to change the default password from "raspberry" to something a little stronger.

System → Boot

This option lets you boot straight to the graphical desktop environment and is set this way by default. If you select CLI, you'll get the command line when you boot up, and you'll have to start the graphical interface manually with the command `startx`.

Display → Overscan

The overscan option is set to enabled at first because some monitors may cut off the edges of the desktop. If you have a black border around your desktop, then you can disable overscan to get the desktop to fill your screen.

Interfaces → SSH

This option turns on the Secure Shell (SSH) server, which will allow you to login to the Raspberry Pi remotely over a network. This is really handy, so you should leave it on.

Performance → GPU Memory

This option allows you to change the allocation of RAM available to the graphics processing unit. The rest of the RAM is left for the CPU to use. It's best to leave the default split for now. If you decide to experiment with 3D graphics or video decoding, you may want to adjust this value in the future.

Performance → Overclock

With this option, you can run the processor at speeds higher than the default operation. This option is not available for Raspberry Pi 3 or 4. For now, it's best to leave this setting alone.

Localisation → Set Keyboard

The default keyboard settings are for a generic keyboard in a UK-style layout. If you want the keys to do what they're labeled to do, you'll definitely want to select a keyboard type and mapping that corresponds to your setup. Luckily, the keyboard list is very robust. Note that your locale settings can affect your keyboard settings as well.

Localisation → Set Locale

If you're outside the UK, you should change your locale to reflect your language and character encoding preferences. The default setting is for UK English with a standard UTF-8 character encoding (en_GB.UTF-8). Select en_US.UTF-8 if you're in the US.

Localisation → SetTimezone

You'll probably want to set this.

When you're done, select OK and you'll be prompted to restart so that the settings can take effect.

If you want to access these settings from the command line, you can use the `raspi-config` tool (see "Configuring Your Pi" on page 17). Type the following at the command line if you want to try that out:

```
sudo raspi-config
```

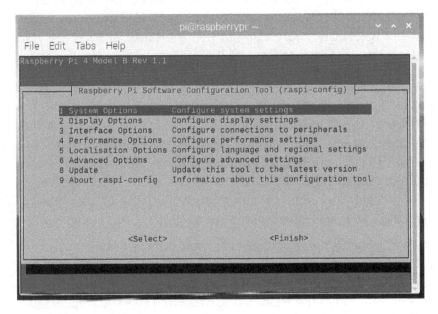

Figure 1-10. *The raspi-config tool when run from the command line*

Getting Online

You've got a few different ways to connect to the internet. If you've got easy access to a router, switch, or Ethernet jack connected to a router, just plug in using a standard Ethernet cable. If you have a Wi-Fi USB dongle or you're using a Raspberry Pi 3 or later, you can connect wirelessly; there's an icon on the task-bar to setup your wireless connection (see Figure 1-10).

If you've got a laptop nearby, or if you're running the Pi in a headless configuration, you can share the Wi-Fi on your laptop with the Pi (Figure 1-11). It is super simple on the Mac: just enable internet Sharing in your Sharing settings, then use an Ethernet cable to connect the Pi and your Mac. In Windows, enable "Allow other network users to connect through this computer's internet connection" in your internet Connection Sharing properties. The Pi should automatically get an IP address when connected and be online.

You will probably need a cross-over cable for a Windows-based PC, but you can use any Ethernet cable on Apple hardware, as it will autodetect the type of cable. (A cross-over cable is a specially-wired type of Ethernet cable —CAT5 or CAT6—that has the "receive" pin on one end connected to the "transmit" pin on the other. This allows two computers to be wired together via their Ethernet ports and be able to talk to each other without having to use an intervening switch or hub in the middle.)

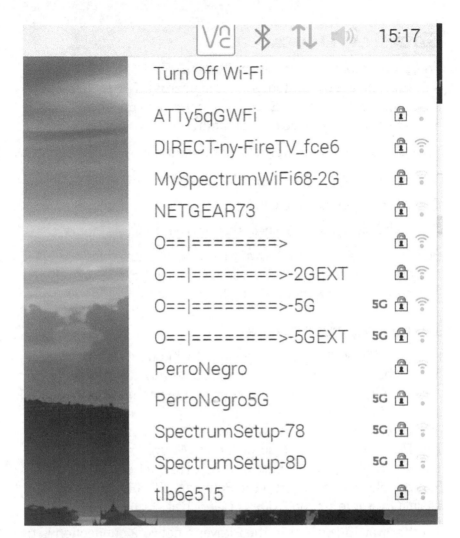

Figure 1-11. *Click on the network icon on the right side of the task bar to select a Wi-Fi network to connect to.*

Shutting Down

There's no power button on the Raspberry Pi (although there is a header for a reset switch on newer boards). The proper way to shut down is through the Shut down command under the taskbar menu with in the desktop environment.

You can also shut down from the command line by typing:

```
pi@raspberrypi: ~ $ sudo shutdown now
```

If you want to restart, you can type

```
pi@raspberrypi: ~ $ sudo shutdown -r now
```

Be sure to do a clean shutdown whenever possible (and don't just pull the plug). In some cases, you can corrupt the SD card if you turn off the power without halting the system.

Running Headless

If you want to work on the Raspberry Pi without plugging in a monitor, keyboard, and mouse, there are some ways to set it up to run *headless*. If all you require is to get into the command line, you can simply hook the Raspberry Pi up to the network and use an SSH client to connect to it (username: `pi`, password: `raspberry`). The SSH utility on Mac or Linux will do; use PuTTY (bit.ly/1sfuf4X) on Windows (or Linux). The SSH server on the Raspberry Pi is enabled by default (run the Raspberry Pi configuration utility again if for some reason it doesn't launch at startup).

Another way to connect to the Pi over a network connection is to start the Virtual Network Computing (VNC) server on the Pi and connect to it using a VNC client. The benefit of this is that you can run a complete working graphical desktop environment in a window on your laptop or desktop. This is a great solution for a portable development environment. The VNC server comes preinstalled on the latest versions of the Raspberry Pi OS; to start and configure it, click the VNC server icon on the top right of the taskbar (Figure 1-12).

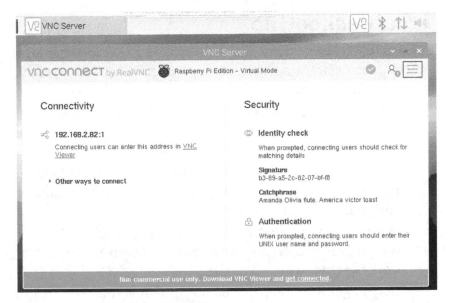

Figure 1-12. *Starting the VNC server*

A third way of logging in to the Pi without a keyboard or monitor is via some pins on the GPIO header. You can use a special cable from FTDI that allows you to connect to that serial port via USB. The FTDI cable has three wires that connect to ground (pin 6), TX (pin 8), and RX (pin 10) on the header. This allows you to login via the serial port, an artifact from olden times when programmers would log into their VAXes and ENIACs via serial cables. Nowadays this type of login is almost never used, as telnet and SSH are available on the vast majority of machines and are much faster communication protocols.

Alternatively, you could use the BUB I from Modern Device, which is a breakout board for the FTDI chip with a prototyping area that allows you to reroute the signals.

Troubleshooting

If things aren't working the way you think they should, there area few common mistakes and missed steps. Be sure to check all of the following suggestions:

- Is the *correct type* of microSD card in the slot, and is it making a good connection?
- Try copying the disk again with another card reader.
- Run a Secure Hash Algorithm (SHA) check sum utility on the disk image and comparing the result to the 40-character hash published on the download page.
- Is the Pi restarting or having intermittent problems? Check your power supply; an underpowered board may seem to work but act flaky.
- Do you get a kernel panic on startup? A kernel panic is the equivalent of Windows' Blue Screen of Death; it's most often caused by a problem with a device on the USB hub. Try unplugging USB devices and restarting.

If that all fails, head over to the Raspberry Pi Hub's troubleshooting page (elinux.org/R-Pi_Troubleshooting) for solutions to all sorts of problems people have had.

Which Board Do You Have?

If you're asking for help in an email or on a forum, it can be helpful to those assisting you if you know exactly what version of the operating system and which board you're using. To find out the OS version, open LXTerminal, and type:

```
cat/proc/version
```

To find your board version, type:

```
cat/proc/cpuinfo
```

Going Further

The Raspberry Pi Hub (elinux.org/RPi_Hub)
   Hosted by elinux.org, this is a massive wiki of information on
   the Pi's hardware and configuration.

List of Verified Peripherals (elinux.org/RPi_VerifiedPeripherals)
   The definitive list of peripherals known to work with the
   Raspberry Pi.

2/Getting Around Linux on the Raspberry Pi

If you want to get the most out of your Raspberry Pi, you'll need to learn a little Linux. This chapter aims to present a whirlwind tour of the operating system and point out the most important tools you'll need today. This should give you enough context and commands to get around the filesystem, and to install packages from the command line or desktop environment.

Raspberry Pi OS comes with the Lightweight X11 Desktop Environment (LXDE) graphical desktop environment installed (Figure 2-1). This is a trimmed-down desktop environment for the X Window System that has been powering the GUIs of Unix and Linux computers since the 1980s. Some of the tools you see on the desktop, such as the LXTerminal shell, are bundled with LXDE.

Running on top of LXDE is Openbox, a window manager that handles the look and feel of windows and menus. If you want to tweak the appearance of your desktop, click the Raspberry menu in the upper left, then choose Preferences→Appearance Settings. Unlike OS X or Windows, it is relatively easy to completely customize your desktop environment or install alternative window managers. Some of the other distributions for Raspberry Pi have different environments tuned for applications like set-top media boxes, phone systems, or network firewalls. See elinux.org/RPi_Distributions and Chapter 3 for more.

As of October 2015's update to the Raspberry Pi OS, the default behavior is to log in and launch the desktop environment immediately after booting. If you find yourself at a prompt for a username and password, the default user is `pi` and the password is `raspberry`. If you find yourself at the text command line (which looks like `pi@raspberrypi:~ $`) and want to launch the desktop environment, just type `startx` and press Enter. We'll cover the command line more in-depth later in this chapter.

Figure 2-1. *The graphical desktop*

The Raspberry Pi software engineers and designers have customized the Pi's Linux distribution and its desktop environment for general-purpose computing, making, and learning. If you browse the programs in the Raspberry menu in the upper-left corner of your screen (Figure 2-2), you'll notice that there are programming environments, office tools, accessories, games, and internet programs preinstalled. Feel free to click around and explore!

Figure 2-2. *The contents of the Raspberry menu*

A few applications that you encounter may be a little different than those in other desktop environments:

The File Manager

If you prefer not to move files around using the command line (more on that in a moment), select the File Manager from the taskbar or within the Raspberry menu→Accessories. You'll be able to browse the filesystem using icons and folders the way you're probably used to doing.

The web browser

The default web browser on the Pi is Chromium (not to be confused with Chrome). Chromium is the open-source web browser upon which Chrome is based and runs quite well on all models of the Pi. If you're used to using Chrome, you probably won't notice much of a difference with Chromium. Advances in the Pi's hardware have made Chromium a perfectly reasonable option for the Pi.

Video and audio

The default video/audio player included in Raspberry Pi OS is VLC. VLC is a powerful, easy-to-use media player that can play pretty much any media file you throw at it.

Wolfram and Mathematica

Bundled with the Raspberry Pi OS (if you use the NOOBS download) are releases of Wolfram language and Mathematica. Mathematica is the front end interface for the Wolfram programming language (no relation to the author). Together they're commonly used for complex computations in math, science, and engineering fields. To see what Wolfram and Mathematica are capable of, the Wolfram Language and System Documentation (bit. ly/1qoitzW) is a great place to start.

Text editor

Previously, the Raspberry Pi came with its own text editor, called Leafpad, but now it's just a simple text editor, similar to gedit on Ubuntu, TextEdit on the Mac, or Notepad on Windows. It's available from the Accessories submenu under the Raspberry icon. You can use nano (nano-editor.org) for editing text files from the command line, as it's preinstalled, as is vim. Emacs (my personal favorite) is not installed but is easy to install afterward (see"Installing New Software" on page 47).

Copy and paste

Copy and paste functions work between applications pretty well, although you may find some oddball programs that aren't consistent. If your mouse has a middle button, you can select text by highlighting it as you normally would (click and drag with the left mouse button) and paste it by pressing the middle button while you have the mouse cursor over the destination window.

The shell

A lot of tasks are going to require you to get to the command line and run commands there. The LXTerminal program provides access to the command line, or shell. It can be launched from the task bar icon or from the Raspberry menu→Accessories. Unfortunately, the standard Debian shortcut for the terminal, Ctrl-Alt-T, does not work on the Pi. The default shell on the Raspberry Pi OS is the Bourne-again shell (bash (bit.ly/1oTTXqW)), which is very common on Linux systems. There's also an alternative called dash (bit.ly/1oTTVzs). You can change shells via the program menu or with the **chsh** command.

Using the Command Line

If it helps, you can think of using the command line as playing a text adventure game, but with the files and the filesystem in place of clues and mazes. If that metaphor doesn't help you, don't worry: all the commands and concepts in this section are standard Linux and are valuable to learn.

Before you start, open up the LXTerminal program (Figure 2-3). There are two tricks that make life much easier in the shell: *autocomplete* and *commandhistory*. Often you will only need to type the first few characters of a command or file name, then hit Tab. The shell will attempt to auto complete the string based on the files in the current directory or programs in commonly used directories (the shell will search for executable programs in places like */bin or/usr/bin/*). If you hit the up arrow on the command line, you'll be able to step back through your command history, which is useful if you mistyped a character in a long string of commands. If you need to go *really* far back in your history, just type **history** in the shell. You'll see a list of commands entered into the shell, going quite a way back, each preceded by a number. If you want to repeat a certain command, just type an exclamation point, followed by the number preceding that command, and press Return.

```
15 python3
16 pip3 install "picamera[array]"
17 mkdir opencv-test
18 cd opencv-test/
19 ls
20 mv Raspberries.jpeg raspberries.jpeg
21 emacs image-display.py
22 sudo apt-get install emacs
23 emacs image-display.py
24 cp image-display.py superimpose.py
25 emacs superimpose.py
26 cp superimpose.py superimposebook.py
27 emacs superimposebook.py
28 cp superimposebook.py superimposesave.py
29 emacs superimposesave.py
30 python3 superimposesave.py
31 emacs superimposesave.py
32 python3 superimposesave.py
33 ls
34 emacs superimposesave.py
35 emacs basic-camera.py
36 exit
37 clear
38 history
pi@raspberrypi:~ $
```

Figure 2-3. *LXTerminal gives you access to the command line (or shell).*

Files and the Filesystem

Table 2-1 shows some of the important directories in the filesystem. Most of these follow the Linux standard of where files should go; a couple are specific to the Raspberry Pi. The */sys* directory is where you can access all of the hardware on the Raspberry Pi.

Table 2-1. *Important directories in the Raspberry Pi OS file system*

Directory	Description
/bin	Programs and commands that all users can run
/boot	All the files needed at boot time
/dev	Special files that represent the devices on your system
/etc	Configuration files
/etc/init.d	Scripts to start up services
/etc/X11	X11 configuration files
/home	User home directories
/home/pi	Home directory for Pi user
/lib	Kernel modules/drivers
/media	Mount points for removable media
/proc	Virtual directory with details about running processes and the OS
/sbin	Programs for system maintenance
/sys	A special directory on the Raspberry Pi that represents the hardware devices
/tmp	Space for programs to create temporary files
/usr	Programs and data usable by all users
/usr/bin	Most of the programs in the operating system reside here
/usr/games	Games (surprise!) No, it's empty by default
/usr/lib	Libraries to support common programs
/usr/local	Software that may be specific to this machine goes here
/usr/sbin	More system administration programs
/usr/share	Supporting files that aren't specific to any process or architecture
/usr/src	Linux is open-source; here's the source!
/var	System logs and spool files
/var/backups	Backup copies of all the most vital system files
/var/cache	Programs such as `apt-get` cache their data here
/var/log	All of the system logs and individual service logs
/var/mail	All user email is stored here, if you're setup to handle email
/var/spool	Data waiting to be processed (e.g., incoming email, print jobs)

You'll see your current directory displayed before the command prompt. In Linux, your home directory has a shorthand notation: the tilde (~). When you open the LXTerminal, you'll be dropped into your home directory, and your prompt will look like this:

```
pi@raspberrypi:~ $
```

Here's an explanation of that prompt:

```
pi@ ❶ raspberrypi: ❷ ~ ❸ $ ❹
```

❶ Your username, pi, followed by the at (@) symbol.

❷ The name of your computer (raspberry pi is the default host name). It can be changed.

❸ The *current working directory* of the shell. You always start in your home directory (~). If you should change directories into the Documents directory, the prompt will change to `pi@raspberrypi:~/Documents $`

❹ This is the *shell prompt*. Any text you type will appear to the right of it. Press Enter or Return to execute each command you type.

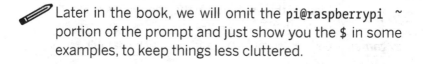 Later in the book, we will omit the `pi@raspberrypi ~` portion of the prompt and just show you the $ in some examples, to keep things less cluttered.

Use the `cd` (change directory) command to move around the file system. The following two commands have the same effect (changing to the home directory) for the Pi user:

```
cd /home/pi/
cd ~
```

In addition, if you just type `cd`, you'll be taken to the home directory. It's a handy shortcut to know.

If the directory path starts with a forward slash, it will be interpreted as an absolute path to the directory. Otherwise, the directory will be considered relative to the current working directory. You can also use `.` and `..` to refer to the current directory and the current directory's parent. For example, to move up to the top of the filesystem:

```
pi@raspberrypi:~ $ cd..
pi@raspberrypi:/home $ cd..
```

You could also get there with the absolute path/:

```
pi@raspberrypi:~ $ cd /
```

Once you've changed to a directory, use the `ls` command to list the files there:

```
pi@raspberrypi:/ $ ls
bin  dev  home  lost+found  mnt  proc  run   selinux  sys  usr
boot etc  lib   media       opt  root  sbin  srv      tmp  var
```

Most commands have additional parameters, or *switches*, that can be used to turn on different behaviors. For example, the -1 switch will produce a more detailed listing, showing file sizes, dates, and permissions:

```
pi@raspberrypi:~ $ ls -l
total 8
drwxr-xr-x2pipi4096Oct1214:26Desktop
drwxrwxr-x2pipi4096Jul2014:07python_games
```

The -a switch will list all files, including invisible ones (invisible file names begin with a dot):

```
pi@raspberrypi:~ $ ls -la
total 80
```

```
drwxr-xr-x  11  pi    pi    4096  Oct 12  14:26  .
drwxr-xr-x  3   root  root  4096  Sep 18  07:48  ..
-rw-------  1   pi    pi    25    Sep 18  09:22  .bash_history
-rw-r--r--  1   pi    pi    220   Sep 18  07:48  .bash_logout
-rw-r--r--  1   pi    pi    3243  Sep 18  07:48  .bashrc
drwxr-xr-x  6   pi    pi    4096  Sep 19  01:19  .cache
drwxr-xr-x  9   pi    pi    4096  Oct 12  12:57  .config
drwx------  3   pi    pi    4096  Sep 18  09:24  .dbus
drwxr-xr-x  2   pi    pi    4096  Oct 12  14:26  Desktop
-rw-r--r--  1   pi    pi    36    Sep 18  09:35  .dmrc
drwx------  2   pi    pi    4096  Sep 18  09:24  .gvfs
drwxr-xr-x  2   pi    pi    4096  Oct 12  12:53  .idlerc
-rw-------  1   pi    pi    35    Sep 18  12:11  .lesshst
drwx------  3   pi    pi    4096  Sep 19  01:19  .local
-rw-r--r--  1   pi    pi    675   Sep 18  07:48  .profile
drwxrwxr-x  2   pi    pi    4096  Jul 20  14:07  python_games
drwx------  4   pi    pi    4096  Oct 12  12:57  .thumbnails
-rw-------  1   pi    pi    56    Sep 18  09:35  .Xauthority
-rw-------  1   pi    pi    300   Oct 12  12:57  .xsession-errors
-rw-------  1   pi    pi    1391  Sep 18  09:35  .xsession-errors.old
```

```
drwxr-xr-x  11  pi    pi    4096  Oct 12  14:26  .
drwxr-xr-x  3   root  root  4096  Sep 18  07:48  ..
-rw-------  1   pi    pi    25    Sep 18  09:22  .bash_history
-rw-r--r--  1   pi    pi    220   Sep 18  07:48  .bash_logout
-rw-r--r--  1   pi    pi    3243  Sep 18  07:48  .bashrc
drwxr-xr-x  6   pi    pi    4096  Sep 19  01:19  .cache
drwxr-xr-x  9   pi    pi    4096  Oct 12  12:57  .config
drwx------  3   pi    pi    4096  Sep 18  09:24  .dbus
drwxr-xr-x  2   pi    pi    4096  Oct 12  14:26  Desktop
-rw-r--r--  1   pi    pi    36    Sep 18  09:35  .dmrc
drwx------  2   pi    pi    4096  Sep 18  09:24  .gvfs
drwxr-xr-x  2   pi    pi    4096  Oct 12  12:53  .idlerc
-rw-------  1   pi    pi    35    Sep 18  12:11  .lesshst
drwx------  3   pi    pi    4096  Sep 19  01:19  .local
-rw-r--r--  1   pi    pi    675   Sep 18  07:48  .profile
drwxrwxr-x  2   pi    pi    4096  Jul 20  14:07  python_games
drwx------  4   pi    pi    4096  Oct 12  12:57  .thumbnails
-rw-------  1   pi    pi    56    Sep 18  09:35  .Xauthority
-rw-------  1   pi    pi    300   Oct 12  12:57  .xsession-errors
-rw-------  1   pi    pi    1391  Sep 18  09:35  .xsession-errors.old
```

Use the mv command to rename a file. The **touch** command can be used to create an empty dummy file:

```
pi@raspberrypi:~ $ touch foo
pi@raspberrypi:~ $ ls
foo   Desktop   python_games
pi@raspberrypi:~ $ mv foo baz
pi@raspberrypi:~ $ ls
baz   Desktop   python_games
```

Remove a file with rm. To remove a directory, you can use rmdir if the directory is empty, or rm -r if it isn't. The -r is a parameter sent to the rm command that indicates it should recursively delete everything in the directory.

---

 rm -r is a bit like nuclear war—it is easy to start, and it does great damage. Make sure you're in exactly the right directory before you set it off.

If you want to find out all the parameters for a particular command, you can read the user manual with the man command (or you can often use the --help option):

```
pi@raspberrypi ~ $ man curl
pi@raspberrypi ~ $ rm --help
```

To make a new directory, use mkdir. To bundle all of the files in a directory into a single file, use the tar command, originally created for tape archives.

You'll find a lot of bundles of files or source code are distributed as tar files, and they're usually also compressed using the gzip command. Try this:

```
pi@raspberrypi ~ $ mkdir myDir
pi@raspberrypi ~ $ cdmy Dir
pi@raspberrypi ~ $ touch foo bar baz
pi@raspberrypi ~ $ cd ..
pi@raspberrypi ~ $ tar -cf myDir.tar myDir
pi@raspberrypi ~ $ gzip myDir.tar
```

You'll now have a *.tar.gz* archive of that directory that can be distributed via email or the internet.

More Linux Commands

One of the reasons that Linux (and Unix) is so successful is that the main design goal was to build a very complicated system out of small, simple modular parts that can be chained together. You'll need to know a little bit about two pieces of this puzzle: *pipes* and *redirection*.

Pipes are simply a way of chaining two programs together so the output of one can serve as the input to another. All Linux programs can read data from *standard input* (often referred to as *stdin*), write data to *standard output* (*stdout*), and throw error messages to *standard error* (*stderr*). A pipe lets you hook up stdout from one program to stdin of another (Figure 2-4). Use the | operator, as in this example:

```
pi@raspberrypi ~ $ ls -la | less
```

In the above example, the output of the ls command is sent to the input of the less program, which prints data one screenful at a time. (Press q to exit the **less** program.)

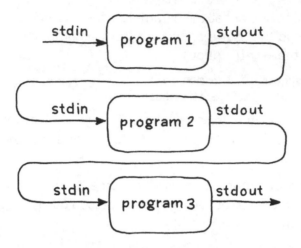

Figure 2-4. *Pipes are a way of chaining smaller programs together to accomplish bigger tasks.*

Now (for something a little more out there) try:

```
pi@raspberrypi ~ $ sudo cat/boot/kernel.img | aplay
```

You may want to turn the volume down a bit first; this command reads the kernel image and spits all of the 1s and 0s at the audioplayer. That's what your kernel sounds like!

Amusing Digression

sudo is a command that stands for "super user do". It is a way for an ordinary computer user, like you, to temporarily gain super user powers on a Linux system. This gives you the ability to change the system greatly, which can do a lot of damage if you're not careful. Treat **sudo** with respect.

In some of the examples later in the book, we'll also be using *redirection*, where a command is executed and the stdout output can be sent to a file. As you'll see later, many things in Linux are treated like ordinary files (such as the Pi's general-purpose input/output pins), so redirection can be quite handy. To redirect output from a program, use the > operator:

```
pi@raspberrypi ~ $ ls > directoryListing.txt
```

Special Control Keys

In addition to the keys for auto complete (Tab) and command history (up arrow) previously mentioned, there are a few other special control keys you'll need in the shell. Here are a few:

Ctrl-C

Kills the running program. May not work with some interactive programs such as text editors.

Ctrl-D

Exits the shell. You must type this at the command prompt by itself (don't type anything after the $ before hitting Ctrl-D).

Ctrl-A

Moves the cursor to the beginning of the line.

Ctrl-E

Moves the cursor to the end of the line.

There are others, but these are the core keyboard shortcuts you'll use every day.

Sometimes you'll want to display the contents of a file on the screen. If it's a text file and you want to read it one screen at a time, use `less`:

```
pi@raspberrypi ~ $ ls >flob.txt
pi@raspberrypi ~ $ less flob.txt
```

(Another option for reading one screen at a time is `more` (I know, I know...). Most power users use `less`, however, as it's a bit more functional than `more` (less is more than more...?).

If you want to just dump the entire contents of a file to standard output, use `cat` (short for concatenate). This can be handy when you want to feed a file into another program or redirect it some-where.

For example, this is the equivalent of copying one file to another with a new name (the second line concatenates the two files first):

```
pi@raspberrypi ~ $ ls >wibble.txt
pi@raspberrypi ~ $ cat wibble.txt > wobble.txt
pi@raspberrypi ~ $ cat wibble.txt wobble.txt > wubble.txt
```

To look at just the last few lines of a file (such as the most recent entry in a log file), use tail (to see the beginning, use head). If you are searching for a string in one or more files, use the venerable program grep:

```
pi@raspberrypi ~ $ grep Puzzle */*
```

grep is a powerful tool because of the rich language of *regular expressions* that was developed for it. Regular expressions can be a bit difficult to read—going into detail of how regular expressions work could fill an entire book—and maybe a major factor in whatever reputation Linux has for being opaque to newcomers.

Processes

Even when the Pi is "doing nothing", its operating system is very busy, maintaining the file system, monitoring the network connection, and many other tasks to keep your Pi humming. Every program on the Pi runs as a separate process; at any particular point, you'll have dozens of processes running. When you first bootup, about 75 processes will start, each one handling a different task or service. To see all these processes, run the top program, which will also display CPU and memory usage. top will show you the processes using the most resources; use the ps command to list all the processes and their ID numbers (see Figure 2-5).

```
File  Edit  Tabs  Help
top - 08:35:37 up 13 min,  2 users,  load average: 0.31, 0.34, 0.35
Tasks: 141 total,   1 running, 140 sleeping,   0 stopped,   0 zombie
%Cpu(s):  1.2 us,   0.2 sy,   0.0 ni, 98.7 id,   0.0 wa,   0.0 hi,   0.0 si,   0.0 st
MiB Mem :   924.2 total,    363.2 free,    158.9 used,    402.1 buff/cache
MiB Swap:   100.0 total,    100.0 free,      0.0 used.    689.8 avail Mem

  PID USER      PR  NI    VIRT    RES    SHR S  %CPU  %MEM     TIME+ COMMAND
 3318 root      20   0   55912  30252  16340 S   2.6   3.2   0:11.33 vncserver-x11-c
 4694 pi        20   0    7864   2688   2172 R   1.3   0.3   0:00.14 top
  506 root      20   0  218772  59120  32728 S   1.0   6.2   0:14.77 Xorg
  109 root      20   0   33232   7220   6260 S   0.7   0.8   0:01.45 systemd-journal
 4649 pi        20   0  273532  56820  42668 S   0.3   6.0   0:03.82 lxterminal
    1 root      20   0   34808   8316   6480 S   0.0   0.9   0:08.31 systemd
    2 root      20   0       0      0      0 S   0.0   0.0   0:00.00 kthreadd
    3 root       0 -20       0      0      0 I   0.0   0.0   0:00.00 rcu_gp
    4 root       0 -20       0      0      0 I   0.0   0.0   0:00.00 rcu_par_gp
    7 root      20   0       0      0      0 I   0.0   0.0   0:00.54 kworker/u8:0-events_+
    8 root       0 -20       0      0      0 I   0.0   0.0   0:00.00 mm_percpu_wq
    9 root      20   0       0      0      0 S   0.0   0.0   0:00.00 rcu_tasks_rude_
   10 root      20   0       0      0      0 S   0.0   0.0   0:00.00 rcu_tasks_trace
   11 root      20   0       0      0      0 S   0.0   0.0   0:03.15 ksoftirqd/0
   12 root      20   0       0      0      0 I   0.0   0.0   0:00.31 rcu_sched
   13 root      rt   0       0      0      0 S   0.0   0.0   0:00.05 migration/0
   14 root      20   0       0      0      0 S   0.0   0.0   0:00.00 cpuhp/0
   15 root      20   0       0      0      0 S   0.0   0.0   0:00.00 cpuhp/1
   16 root      rt   0       0      0      0 S   0.0   0.0   0:00.05 migration/1
   17 root      20   0       0      0      0 S   0.0   0.0   0:00.09 ksoftirqd/1
```

Figure 2-5. *The* **top** *command shows all the processes running, as well as CPU and memory usage.*

Try:

> pi@raspberrypi ~ $ **ps aux | less**

Sometimes you may want to kill a rogue or unresponsive process. To do that, use ps to find its ID, then use kill to stop it:

> pi@raspberrypi ~ $ **kill 95689**

In the case of some system processes, you won't have permission to kill it (though you can circumvent this with sudo, covered in the next section).

Sudo and Permissions

Linux is a multiuser operating system; the general rule is that everyone owns their own files and can create, modify, and delete them within their own space on the filesystem. The root (or super) user can change any file in the filesystem, which is why it is good practice to not login as root on a day-to-day basis. There is a saying among Linux users: "Only noobs log in as root."

There are some tools like sudo ("su–peruser–do") which allow users to act like superusers for performing tasks, like installing software without the dangers (and responsibilities) of being logged in as root. You'll be using sudo a lot when interacting with hardware directly, or when changing system-wide configurations, such as when you're installing software.

 As the pi user, there's not much damage you can do to the system. As superuser, you can wreak havoc, accidentally or by design. Be careful when using sudo, especially when moving or deleting files. Of course, if things go badly, you can always make a new SD card image (see Appendix A).

Each file belongs to one user and one group. Use chown and chgrp to change the file's owner or group. You must be root to use either:

```
pi@raspberrypi ~ $ sudo chown pi garply.txt
pi@raspberrypi ~ $ sudo chgrp staff plugh.txt
```

Each file also has a set of *permissions* that show whether a file can be read, written, or executed. These permissions can be set for the owner of the file, the group, or for everyone (see Figure 2-6).

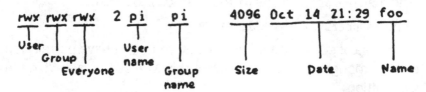

Figure 2-6. *File permissions for owner, group, and everyone*

You set the individual permissions with the chmod command. The switches for chmod are summarized in Table 2-2.

Table 2-2. *The switches that can be used with* **chmod**

u	User
g	Group
o	Others not in the group
a	All/everyone
r	Read permission
w	Write permission
x	Execute permission
+	Add permission
-	Remove permission

Here are a few examples of how you can combine these switches:

```
chmodu+rwx,g-rwx,o-rwx wibble.txt  ❶
chmodg+wx wobble.txt  ❷
chmod -rw, +r wubble.txt  ❸
```

❶ Allow only the user to read, write, and execute.

❷ Add permission to write and execute to the entire group. Make read-only for everyone.

❸ The only thing protecting your user space and files from other people is your password, so you better choose a strong one. Use the `passwd` command to change it, especially if you're putting your Pi on a network.

The Network

Once you're on a network, there are several Linux utilities that you'll be using regularly. When you're troubleshooting an internet connection, use ifconfig, which displays all of your network interfaces and the IP addresses associated with them (see Figure 2-7).

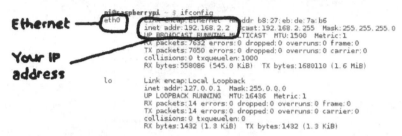

Figure 2-7. *The* ifconfig *command gives you information about all of your network interfaces.*

The ping command is actually the most basic tool for trouble-shooting network connections. Ping acts like sonar on the internet— it sends out a simple pulse to a specific network address and measures the time it takes for the pulse to return, to test whether there is a two-way connection between two IP addresses on the network or internet. Note that many websites block ping traffic, so you may need to ping multiple sites to accurately test a connection:

```
ping google.com
ping opendns.com
ping cloudflare.com
ping netscape.com
```

To log into another computer remotely (and securely, with encrypted passwords), you can use the Secure Shell (SSH). The computer on the remote side needs to be running an SSH server for this to work, but the SSH client comes built into the Raspberry Pi OS and it's easy to enable the SSH server as well. In fact, this is a great way to work on your Raspberry Pi without a monitor or keyboard, as discussed in "Running Headless" on page 24.

Related to SSH is the sftp program, which allows you to securely transfer files from one computer to another. Rounding out the set is scp, which you can use to copy files from one computer to another over a network or the internet. The key to all of these tools is that they use the Secure Sockets Layer (SSL) to transfer files with encrypted login information. These tools are all standard stalwart Linux tools.

/etc

The /etc directory holds all of the system-wide configuration files and startup scripts. When you ran the configuration scripts the first time you started up, you were changing values in various files in the /etc directory. You'll need to invoke superuser powers with sudo to edit files in /etc; if you come across some tutorial that tells you to edit a configuration file, use a text editor to edit and launch it with sudo:

```
pi@raspberrypi ~ $ sudo nano /etc/hosts
```

Setting the Date and Time

A typical laptop or desktop will have additional hardware and a backup battery (usually a coin cell) to save the current time and date. The Raspberry Pi does not, but the Raspberry Pi OS is config-ured to automatically synchronize its time and date with a Network Time Protocol (NTP) server when plugged into a network.

Having the correct time can be important for some applications (see the example in Chapter 6 using cron to control a lamp). To set the time and date manually, use the date program:

```
$ sudo date --set="Sun Nov 20 1:55:16 EST 2022"
```

If the time was set automatically via the internet with NTP, you may want to update your time zone. To do this, go to the International-isation Settings within the raspi-config utility (see "BootingUp" on page 16).

Installing New Software

One of the areas where Linux completely trounces other operating systems is in software package management. *Package managers* handle the downloading and installation of software, and they

automatically handle downloading and installing dependencies—
the other software the package relies upon. The package manager
keeps it all straight, and the package managers on Linux are
remarkably robust.

The Raspberry Pi OS comes with a pretty minimal set of software,
so you will soon want to start downloading and installing new
programs. The examples in this book will all use the command line
for this task because it is the most flexible and quickest way of
installing software.

The program apt-get with the install option is used to download
software. apt-get will even download any extra software or libraries
required so you don't have to go hunting around for dependencies.
The software has to be installed with superuser permissions, so
always use sudo. For example, this command installs the *excellent*
Emacs text editor:

    pi@raspberrypi ~ $ sudo apt-get install emacs

 Taking a Screenshot

> One of the first things we needed to figure out when
> writing this book was how to take screenshots on the
> Pi. Pre-installed in the Raspberry Pi OS is a program
> called scrot (an abbreviation for SCReen-shOT). To take
> a screenshot, just type scrot on the command line. A
> *.png* file will be saved in your home directory. Scrot is
> powerful and has a lot of command-line options; type
> scrot -h to see a short guide on how to use it. Another
> way to invoke scrot is to press your keyboard's *Print
> Screen* button.

Sound in Linux

Raspberry Pi has the built-in capability to play sound. This makes
it a popular platform for DIY projects that play sound effects or
stream music from the internet. The Pi uses the *Advanced Linux
Sound Architecture*, or ALSA, for low-level control of audio devices.

You can test it with a pre-loaded sound file and the sound playback utility aplay:

```
$ aplay /usr/share/scratch/Media/Sounds/Human/PartyNoise.wav
```

If you're using an HDMI display, by default, the sound will be played through the HDMI device.

To adjust the volume output, run alsamixer and use your arrow keys to change the gain.

You're not limited to the onboard sound; you can also add USB audio devices. Many USB audio devices even have an audio input so that you can use your Raspberry Pi to record audio in addition to playing it.

The easiest way to enable a particular audio device is to open VLC from the Sound & Video menu. Once VLC is open, click on Audio at the top of the window and select your preferred audio device.

Upgrading Your Firmware

Some of your Raspberry Pi's *firmware* is stored on the SD card and includes much of the low-level instructions that need to be executed before the boot process is handed over to your operating system.

While it's typically not necessary, if you run into some strange behavior, you may want to try to update the firmware on your SD card. With an Internet-connected Raspberry Pi, this is very easy to do:

```
$ sudo rpi-update
```

If you'd like to see what's being updated when you run the utility, you can review the latest changes in Raspberry Pi's firmware repository on GitHub (github.com/raspberrypi/firmware).

To view what version of the Raspberry Pi firmware you currently have, run vcgencmd version:

```
$ vcgencmd version
Jan 27 2021 22:26:53
Copyright (c) 2012 Broadcom
version c156d00b148c30a3ba28ec376c9c01e95a77d6d5 (clean)
(release) (start)
```

Going Further

There's much more to Linux and many places to continue learning about it. Some good starting points are:

Linux Pocket Guide
by Daniel J. Barrett
Handy as a quick reference.

Linux in a Nutshell
by Ellen Siever, Stephen Figgins, Robert Love, and Arnold Robbins
More detailed, but still a quick reference guide.

The Debian Wiki (wiki.debian.org/FrontPage)
The Raspberry Pi OS is based on Debian, so a lot of the info on the Debian wiki applies to the Pi as well.

The Jargon File (catb.org/jargon)
by Eric S. Raymond
Also published as the *New Hacker's Dictionary*, this collection of definitions and stories is required reading on the Unix/Linux subculture.

3/Other Operating Systems and Linux Distributions

The stock Raspberry Pi OS is great for general-purpose computing, but sometimes you may want to tailor the Pi to a specific purpose, like making it a standalone media center or a guitar effects pedal. The Linux ecosystem is rich in software for every imaginable application. A number of folks have spent the time to bundle all the right software together so you don't have to do it. This chapter will highlight just a few of the more specialized Linux distributions and other operating systems to get you started.

When talking about "Linux distributions," we're usually talking about three things together:

- The Linux kernel and drivers
- Preinstalled software for a particular application
- Special configuration tools or tools preconfigured for a particular task (e.g., to boot up into a particular program)

As you saw in Chapter 1, there are essentially four popular general-purpose distributions. You'll find the first three in the NOOBS installer:

Raspberry Pi OS (raspbian.org)
The recommended distribution from the Foundation to start with; based on Debian. If you're not sure which distribution to choose, this is the one for you.

Arch Linux (www.archlinux.org)
Arch Linux specifically targets ARM-based computers, so they supported the Pi very early on.

Pidora (pidora.ca)
Pidora is a version of the Fedora distribution tuned for the Pi.

Ubuntu MATE (ubuntu-mate.org)
Ubuntu MATE is a version of the very popular Ubuntu distribution of Linux. It has a slimmed-down desktop environment that works rather well on Raspberry Pi 2 and subsequent versions. It won't work on previous versions of Raspberry Pi because Ubuntu only supports ARMv7 and later.

Here are a few other interesting specialized Linux distributions and operating systems.

Distributions for Home Theater

A long-time favorite operating system for home theater is XBMC, which began as a media center project to run on the Xbox game console. Over the years, however, XBMC morphed to become a more general entertainment center platform and happens to work very well on the Pi. In the summer of 2014, the XBMC Foundation renamed the software Kodi to bring the evolution of the project into focus, because it doesn't even run on the newer Xbox versions. There are a couple of Pi distributions that make it easy to put Kodi in your living room:

OSMC (osmc.tv)
Formerly called Raspbmc, the OSMC distribution is based on

Debian and Kodi. It has great support for Raspberry Pi and can be installed from NOOBS (Figure3-1).

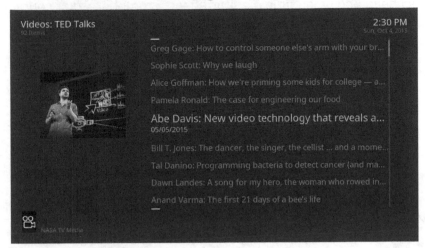

Figure 3-1. *OSMC's interface*

OpenELEC (openelec.tv)

The Open Embedded Linux Entertainment Center is a pared-down version of Kodi that may appeal to more ascetic Pi users (Figure 3-2).

Figure 3-2. *The main menu for The Open Embedded Linux Entertainment Center (OpenELEC)*

Distributions for Music

It's cheap and it can fit in a guitar effects stompbox, so of course the electronic music world has been excited about the Pi since its release. Here are some examples:

Satellite CCRMA (stanford.io/1riPJsE)
This distribution from Stanford's Center for Computer Research in Music and Acoustics (CCRMA) is geared toward embedded musical instruments and art installations, as well as effects pedals. The original rationale is described in Edgar Berdahl and Wendy Ju's paper "Satellite CCRMA: A Musical Interaction and Sound Synthesis Platform" (bit.ly/1qol7Wo).

Volumio (volumio.org)
A music player for audiophiles. This project evolved from Raspy-Fi (www.hifiberry.com/hbdigi).

PiCore Player (www.picoreplayer.org)
This is a full-fledged music player distribution for almost all models of the Pi, enabling you to play your Spotify, Tidal, or other streaming music service playlists on your Pi, as well as local music playlists on a local drive.

PiMusic Box (www.pimusicbox.com)
Another music player distribution for local and streaming playlists. Like PiCore Player, it's exceedingly small, which means it can run on any version of Pi, even the Zero W.

Also on the music front, you may want to check out the Sun Vox modular music platform for the Pi (www.warmplace.ru/soft/sunvox).

Although it's not limited to the Raspberry Pi platform, there is also a fascinating program called Sonic Pi (sonic-pi.net) that music-loving programmers might want to check out. It enables you to create music in real-time by writing coding commands, and has been used in environments as diverse as teaching and ambient music in nightclubs.

Retro Computing and Retro Gaming

The Pi was inspired by the inexpensive personal computers of the 1980s, so it seems fitting that there are a number of distributions aimed at nostalgic retro computing or gaming:

RISCOS (www.riscosopen.org/content)
Boots straight into BASIC!

Retropie (retropie.org.uk)
An SD card image and GPIO hardware board that makes it easier to build retro gaming consoles.

PiPlay (piplay.org)
A prebuilt distribution for gaming and emulation based on MAME (formerly PiMAME).

 It's not an OS, but if you're into retro text adventures, try Frotz:

```
sudo apt-get install frotz
```

Internet of Things

The *Internet of Things* or *IoT* describes the realm of devices connected to the internet. These can be thermostats, body weight scales, and doorbells that can be accessed remotely via the Weborina mobile app. The realm of IoT extends beyond the home as well. Large companies put their equipment online to monitor their assets, which may be spread all over the globe.

Because of Raspberry Pi's low cost and connectivity, it makes a great choice for experimentation with the Internet of Things, especially the Pi Zero W. While you could use Raspberry Pi OS for this (in fact, see "Connecting the Web to the Real World" on page 181), here are a few operating systems that are geared toward IoT.

UbuntuCore (ubuntu.com/core)

Available for the Raspberry Pi models 2, 3, and 4, UbuntuCore is a Linux distribution aimed at devices and cloudservers. It's a bare bones operating system that includes methods for being very specific about which versions of applications are installed to support your project. It's meant to be leaner, faster, more reliable, and more secure. It was originally designed to be just for embedded applications, but it has grown along with the Pi's capabilities, and you can now install a full Ubuntu server on your Pi!

Windows 10 IoT Core (dev.windows.com/en-us/iot)

With the release of Windows 10, Microsoft doubled down on IoT with the introduction of the Windows 10 IoT Core. While it doesn't have the full Windows desktop environment, developers can now deploy universal Windows applications to Raspberry Pi 2 and 3. (Unfortunately, as I write this, IoT Core does not run on the Pi 4.) If you already have experience developing Windows applications, trying out Windows 10 IoT Core on the Raspberry Pi is a no-brainer. If you're interested in getting started, Microsoft provides excellent getting-started documentation.

Other Useful Distributions

Here are a few other distributions of note:

Qt on Pi (www.qt.io)
An OS bundle aimed at developers of standalone "single-purpose appliances"using the QtGUI framework.

Web kiosk (www.binaryemotions.com)
For making Internet kiosks and digital signage.

Openwrt (openwrt.org)
Turn your Pi into a powerful router with this open-source router platform.

OctoPrint (octoprint.org)
With OctoPrint, you can control your 3D printer via your network. Just connect it to a Raspberry Pi and boot from the OctoPiSD card image.

Going Further

List of Linux distributions that work with Raspberry Pi
(bit.ly/1BW4SGp)
This is the definitive list at the Raspberry Pi Hub.

Raspberry Pi Downloads
(www.raspberrypi.org/downloads)
The downloads section of the Raspberry Pi Foundation website will have a listing of notable operating systems and distributions.

4/Python on the Pi

Python is a great first programming language; it's clear and easy to get up and running. More importantly, there are a lot of other users to share code with and ask questions.

Guido van Rossum released Python in 1991, and very early on recognized its usefulness as a first language for computing. In 1999, van Rossum put together a widely read proposal called "Computer Programming for Everybody" (www.python.org/doc/essays/cp4e) that laid out a vision for an ambitious program to teach programming in elementary and secondary schools using Python. More than two decades later, it's actually happening, due in part to the incredible popularity and use of the Raspberry Pi in the classroom.

Python is an *interpreted language*, which means that you can write a program or script and execute it directly, rather than compiling it into machine code. Interpreted languages are a bit quicker to program with, and they give you a few side benefits. For example, in Python you don't have to explicitly tell the computer whether a variable is a number, a list, or a string; the interpreter figures out the data types when you execute the script.

The Python interpreter can be run in two ways: as an interactive shell to execute individual commands, or as a command-line program to execute standalone scripts. The integrated development environment (IDE) often bundled with Python is called IDLE. The latest version of the Raspberry Pi OS no longer comes with IDLE, but instead with Geany and Thonny. Both of these IDEs are a bit more robust, in my opinion, than IDLE, which is more of a simple command line environment.

The Python Version Conundrum

In your experiments on the Pi, you may discover that there are two versions of Python installed on the Pi. This is common practice (though a bit confusing). As of this writing, Python 3 is the newest version of the language, and the only version in active development—support of Python v2.7 stopped in January of 2021. Unfortunately, changes made to the language between versions 2 and 3 made the latter not backward compatible. Even though Python 3 has been around for years, it has taken awhile for it to be widely adopted, and lots of user-contributed packages have still not been upgraded to Python 3. Things get even more confusing when you search the Python documentation; make sure you're looking at the right help file for the version you're working in!

You can explicitly run Python 3 with:

```
python3
```

The examples in this book will work with Python 2.7 or 3.X, unless otherwise noted, but will be written exclusively in Python 3 syntax (using opening and closing parentheses in print statements, for instance.)

Hello, Python

The best way to start learning Python is to jump right in. Although you can use any text editor to start scripting, we'll begin by using the IDLE 3 application. To run IDLE, click the desktop menu in the lower left, and choose Programming→Python (IDLE 3).

When IDLE opens you'll see a window with the interactive shell. The triple chevron (>>>) is the interactive prompt; when you see the prompt, it means the interpreter is waiting for your commands. At the prompt, type the following:

```
>>> print("SalutonMondo!")
```

Hit Enter or Return. Python executes the statement you just typed, and you'll see the result in the shell window. You can use the shell as a kind of calculator to test out statements or calculations. Try this:

```
>>> 3+4+5
12
```

Think of the statements executed in the interactive shell as a program that you're running one line at a time. You can even setup variables or import modules:

```
>>> import math
>>> (1 + math.sqrt(5)) / 2
1.618033988749895
```

The import command makes all of Python's math functions available to your program (more about modules in "Objects and Modules" on page 65). To set up a variable, use the assignment operator(=):

```
>>> import math
>>> radius = 20
>>> radius *2* math.pi
125.66370614359173
```

If you want to clear all variables and start in a fresh state, select Shell → Restart Shell from the menu to start over. You can also use the interactive shell to get information about how to use a particular statement, module, or other Python topics with the help() command:

```
help ("print")
```

To get a listing of all of the topics available, try:

```
help ("topics")
help ("keywords")
help ("modules")
```

The Python interpreter is good for testing statements or simple operations, but you will often want to run your Python script as you would a standalone application. To start a new Python program, select File → New Window, and IDLE will give you a script editing window (see Figure 4-1).

Try typing a line of code and selecting Run→Run Module. You'll get a warning that "Source Must Be Saved OK To Save?". Save your script in your home directory as *SalutonMondo.py* and you'll see it execute in the shell.

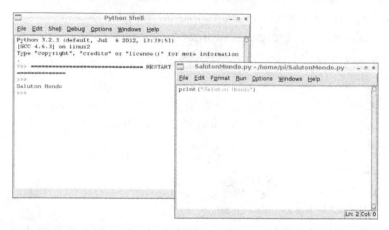

Figure 4-1. *The IDLE interactive shell (left) and an editor window (right)*

Sometimes you may not want to use the IDLE environment. To run a script from the command line, open up the Terminal and type:

```
python3 SalutonMondo.py
```

That's really all of the basic mechanics you need to know to get up and running with the environment. Next, you'll need to start learning the language.

A Bit More Python

If you're coming to Python from the Arduino world, you're used to writing programs (known as *sketches* in Arduino, but often called *scripts* in Python) in a setup/loop format, where setup() is a function run once and loop() is a function that executes over and over. The following example shows how to achieve this in Python. Select New Window from the shell in IDLE 3 and type the following:

```python
# Setup
n=0
# Loop
while True:
    n = n + 1
    # The % is the modulo operator
    if ((n % 2) == 0):
    print(n)
```

Select Run Module, and give your script a name (such as *Count-Evens. py*). As it runs, you should see all even integers printed (press Ctrl-C to interrupt the output, because otherwise, it will go on forever).

You could also implement this using a `for` loop that just counts the first 100 even integers:

```
for n in range (0, 100):
    if ((n % 2) == 0):
        print(n)
```

 In the preceding example, you may not notice that each level of indentation is four spaces, not a tab (but you can press Tab in IDLE, and it will dutifully insert four spaces for you). Indentation has structural meaning in Python, similar to code blocks situated within curly braces in C or Javascript.

This is one of the big stumbling blocks for beginners, especially when copying and pasting code. Still, we feel that the mandatory use of white space makes Python a fairly readable language. See the Style Guide for Python Code (bit.ly/1ne4x05) for tips on writing readable code.

It is important to watch your white space; Python is a highly structured language where the white space determines the structure. In the next example, everything indented one level below the `loop()` function is considered part of that function. The end of a loop is determined by where the indentation moves up a level (or end of the file). This differs from languages like C that delimit blocks of code with brackets or other markers.

Use functions to put chunks of code into a code block that can be called from other places in your script. To rewrite the previous example with functions, do the following (when you go to run this sketch, save it as *CountEvens.py*):

```
# Declare global variables
n=0  ❶
# Setup function
def setup():  ❷
    global n
n = 100
def loop():  ❸
    global n
    n = n + 1
    if ((n % 2) == 0):
    print(n)
# Main  ❹
setup()
while True:
    loop()
```

In this example, the output will be every even number from 102 on. Here's how it works:

❶ First, the variable n is defined as a global variable that can be used in any block in the script.

❷ Here, the setup() function is defined (but not yet executed).

❸ Similarly, here's the definition of the loop()f function.

❹ In the main codeblock, setup() is called once, then loop() is called forever.

The use of the global keyword in the first line of each function is important; it tells the interpreter to use the global variable *n* rather than create a second (local, or private to that function) *n* variable usable only in the function.

This tutorial is too short to be a complete Python reference. To really learn the language, you may want to start with *Learn Python the Hard Way* (learnpythonthehardway.org), *Think Python*, or the *Python Pocket Reference*. The rest of this chapter will give you enough context to get up and running and will map out the basic features and modules available in Python.

Command Line Versus IDLE

One thing you will notice is that the output of IDLE is very slow when running example code that prints to the shell. To get an idea of just how slow it is, keep IDLE open, and start a new terminal alongside it. In IDLE, run the CountEvens script using Run Module; it will need the headstart. Then type the following at the terminal's command line prompt:

```
python3 CountEvens.py
```

You'll quickly get an idea of the overhead from using the IDE on the fairly limited resources of the Pi. The examples later in the book will all be executed from the command line, but IDLE can still be used as an editor if you like.

Objects and Modules

You'll need to understand the basic syntax of dealing with objects and modules to get through the examples in this book. Python is a clean language, with just 35 reserved *keywords* (see Table 4-1). These keywords are the core part of the language that let you structure and control the flow of activity in your script. Pretty much everything that isn't a keyword can be considered an *object*. An object is a combination of data and behaviors that has a name. You can change an object's data, retrieve information from it, and even manipulate other objects.

Table 4-1. *Python has just 35 reserved keywords.*

Conditionals	Loops	Built-in functions	Classes, modules, functions	Error handling
if	for	print	class	try
else	in	pass	def	def
elif	while	del	global	finally
not	break	lambda	break	raise
or	as		nonlocal	assert
and	continue		yield	with
is			import	
True			return	
False			from	
None				

In Python, strings, lists, functions, modules, and even numbers are objects. A Python object can be thought of as an encapsulated collection of attributes and methods. (For those that are unsure, attributes are also sometimes referred to as variables, and methods are also referred to as functions. In the software world, these terms are often interchangeable, particularly methods and functions.) You get access to these attributes and methods using a simple dot syntax. For example, type this at the interactive shell prompt to set up a string object and call the method that tells it to capitalize itself:

```
>>> myString = "quux"
>>> myString.capitalize()
'Quux'
```

Or use **reverse()** to rearrange a list in reverse order:

```
>>> myList = ['a', 'man', 'a', 'plan', 'a', 'canal']
>>> myList.reverse()
>>> print(myList)
['canal', 'a', 'plan', 'a', 'man', 'a']
```

Both string and list are built-in modules of the *standard library*, which are available from any Python program. In each case, the string and list modules have a defined bunch of functions for dealing with strings and lists, including `capitalize()` and `reverse()`.

Some of the standard library modules are not built-in, and you need to explicitly tell the Pi you're going to use them. Do that with the `import` command. For example, to use the `time` module from the standard library to gain access to helpful functions for dealing with timing and time stamps, use:

```
import time
```

You may also see `import as` used to rename the module in your program:

```
import time as myTime
```

Or `from import` used to load only select functions from a module:

```
from time import clock
```

Here's a short example of a Python script using the `time` and `date time` modules from the standard library to print the current time once every second:

```
from datetime import date time
from time import sleep
while True:
    now = str(datetime.now())
    print(now)
    sleep(1)
```

The `sleep` function stops the execution of the program for one second. One thing you will notice after running this code is that the displayed time will drift a bit from reality.

That's for two reasons:

- The code doesn't take into account the amount of time it takes to calculate the current time.
- Other processes are sharing the Pi's CPU, and may take cycles away from your program's execution. This is an important thing

to remember: when programming on the Raspberry Pi, you are not executing in a *real-time environment*.

If you're using the `sleep()` function, you'll find that it is accurate to within about 5ms on the Pi. However, you can read that to also mean that the `sleep()` function is accurate to about 5ms, but no more than that. If you need to pause in your program for a shorter length of time than 5ms, you may need to consider using a real-time operating system or a microcontroller for your application, as the Pi is unlikely to be accurate to within those specs.

Next, let's modify the example to open a text file and periodically log some data to it. In Python, everything is a string when handling text files.

Use the `str()` function to convert numbers to strings (and `int()` to change numeric strings back to an integer. This is referred to as *casting* a variable):

```
from datetime import datetime
from time import sleep
import random

log = open ("log.txt", "w")

for i in range(5):
    now = str(datetime.now())
    # Generate some random data in the range 0-1024
    data = random.randint(0, 1024)
    log.write(now + " " + str(data) + "\n")
    print(".")
    sleep(.9)
log.flush()
log.close()
```

 In a real data-logging application, you'll want to make sure you've got the correct date and time setup on your Raspberry Pi, as described in "Setting the Date and Time" on page 47.

Here's another example (*ReadFile.py*) that reads in a file name as an argument from the command line (run it from the shell with python3 ReadFile.py *filename*). The program opens the file, reads each line as a string, and prints it. Note that print() acts like println() does in other languages; it adds a new line to the string that is printed. The end argument to print() suppresses the new line:

```python
# Open and read a file from command-line argument
import sys

if (len(sys.argv) != 2):
    print("Usage: python3 ReadFile.py filename")
    sys.exit()

scriptname = sys.argv[0]
filename = sys.argv[1]

file = open(filename, "r")
lines = file.readlines()
file.close()

for line in lines:
    print(line,end = '')
```

Even More Modules

One of the reasons Python is so popular is that there are a great number of user-contributed modules that build on the standard library. The Python Package Index (PyPI) (pypi.org) is the definitive list of packages (or modules) available for the language. Some of the more popular modules that are particularly useful on the Raspberry Pi are shown in Table 4-2. You'll be using some of these modules later on, especially the GPIO module to access the general inputs and outputs of the Raspberry Pi.

Table 4-2. *Some packages of particular interest to Pi users*

Module	Description	URL	Package name
RPi.GPIO	Access to GPIO pins	sourceforge.net/ projects/raspber- ry-gpio-python	python- rpi.gpio
GPIOzero	Simplified access to GPIO pins	gpiozero.readthedocs. org(https://gpiozero. readthe docs.org)	python- gpio-zero
Pygame	Gaming frame work	pygame.org	python- pygame
OpenCV	Easy API for Computer Vision	opencv.org	opencv-python
SciPy	Scientific com- puting	www.scipy.org	python-scipy
NumPy	The numerical underpinnings of Scipy	numpy.scipy.org	python-numpy
Flask	Microframework for web develop- ment	flask.palletsprojects. com	python-flask
Feed parser	Atom and RSS feedparser	pypi.python.org/pypi/ feedparser	No package
Requests	"HTTP for Humans"	docs.python-requests. org	python- requests
PIL	Image processing	www.pythonware.com/ products/pil	python- imaging
wxPython	GUI framework	wxpython.org	python- wxgtk2.8
pySerial	Access to the serial port	github.com/pyserial/ pyserial	python- serial
PyUSB	FTDI-USB interface	bleyer.org/pyusb	No package

To use one of these modules, you'll need to download the code, configure the package, and install it. The NumPy module, for example, can be installed as follows:

```
sudo apt install python-numpy
```

If a package has been bundled by its creator using the standard approach to bundling modules (with Python's distutils tool), all you need to do is download the package, uncompress it, and type:

```
python3 setup.py install
```

Easy Module Installs with Pip

Many modules can be installed with apt. You may also want to look at the Pip package installer (pip.pypa.io), a tool that makes it quite easy to install packages from the PyPI. Install Pip using apt:

```
sudo apt install python3-pip
```

Then you can install most modules using Pip to manage the downloads and dependencies. For example:

```
pip3 install flask
```

In later chapters, you'll use application-specific modules extensively, but here's one example that shows how powerful some modules can be. The Feedparser module is a universal parser that lets you grab RSS or Atom feeds and easily access the content. Because most streams of information on the Web have RSS or Atom output, Feedparser is one of several ways to get your Raspberry Pi hooked into the Internet of Things.

First, install the Feedparser module using Pip (see "Easy Module Installs with Pip" above):

```
pip3 install --user feedparser
```

Feedparser gives you the ability to download and parse RSS feeds from within your Python code—hence the name.

To use it, simply give the parse function the URL of an RSS feed. Feedparser will fetch the XML of the feed and parse it, and turn it into a special list data structure called a *dictionary*. A Python dictionary is a list of key/value pairs, sometimes called a *hash* or *associative array*. The parsed feed is a dictionary, and the parsed items in the feed are also a dictionary, as shown in this example, which grabs the current weather in Providence, RI, from weather.gov:

```
import feedparser

feed_url = "http://w1.weather.gov/xml/current_obs/KPVD.
rss"
feed = feedparser.parse(feed_url)
RSSitems = feed["items"]
for item in RSSitems:
    weather = item["title"]
    print(weather)
```

Launching Other Programs from Python

Python makes it fairly easy to trigger other programs on your Pi with the sys.subprocess module. Try the following:

```
from datetime import datetime
from time import sleep
import subprocess
for count in range(0,  60):
filename = str(datetime.now()) + ".jpg"
subprocess.call(["fswebcam", filename])
sleep(60)
```

This is a simple time-lapse script that will snap a photo from a webcam once a minute for an hour. Cameras are covered in greater detail in Chapter 9, but this program should work with a USB camera connected and the fswebcam program installed; run this command first to install it:

```
sudo apt install fswebcam
```

The subprocess.call function takes a list of strings, concatenates them together (with spaces between), and attempts to execute the program specified. If it is a valid program, the Raspberry Pi will spawn a separate process for it, which will run alongside the

Python script. The command in the preceding example is the same as if you typed the following at a terminal prompt:

```
fswebcam20220812.jpg
```

To add more parameters to the program invocation, just add strings to the list in the `subprocess.call` function:

```
subprocess.call(["fswebcam", "-r", "1280x720", filename])
```

This will send the `fswebcam` program a parameter to change the resolution of the snapped image.

Running a Python Script Automatically at Startup

Because the Pi can act as a standalone appliance, a common question is how to launch a Python script automatically when the Pi boots up. The answer is to add an entry to the */etc/ rc.local* file, which is used for exactly this purpose in the Linux world. Just edit the file:

```
sudo nano /etc/rc.local
```

and add a command to execute your script between the commented section and `exit0`. Something like:

```
/usr/bin/python3/home/pi/foo.py&
```

The ampersand at the end will run the script as a background process, which will allow all the other services of the Pi to continue booting up. Note also that you are using the full path name of the Python executable, which may be necessary if your Pi has several users.

Troubleshooting Errors

Inevitably, you'll run into trouble with your code, and you'll need to track down and squash a bug. The IDLE interactive mode can be your friend; the Debug menu provides several tools that will help you understand how your code is actually executing. You also have the option of seeing all your variables and stepping through the execution line by line.

Syntax errors are the easiest to deal with; usually, this is just a typo or a misunderstood aspect of the language. *Semantic errors*—where the program is well-formed but doesn't perform as expected—can be harder to figure out. That's where the debugger can really help unwind a tricky bug. Effective debugging takes years to learn, but here is a quick cheat sheet of things to check when programming the Pi in Python:

* Use `print()` to show when the program gets to a particular point.
* Use `print()` to show the values of variables as the program executes.
* Double-check whitespace to make sure blocks are defined the way you think they are.
* When debugging syntax errors, remember that the actual error may have been introduced well before the interpreter reports it.
* Double-check all of your global and local variables.
* Check for matching parentheses.
* Make sure the order of operations is correct in calculations; insert parentheses if you're not sure. For example, `3 + 4 * 2` and `(3 + 4) * 2` yield different results.

After you're comfortable and experienced with Python, you may want to look at the `code` and `logging` modules for more debugging tools.

Going Further

There is a lot more to Python, and here are some resources that you'll find useful:

Think Python
by Allen Downey
This is a clear and fairly concise approach to programming (that happens to use Python).

Python Pocket Reference
by Mark Lutz
Because sometimes flipping through a book is better than clicking through a dozen Stack Overflow posts.

Stack Overflow (stackoverflow.com)
That said, Stack Overflow is an excellent source of collective knowledge. It works particularly well if you're searching for a specific solution or error message; chances are someone else has had the same problem and posted it here.

Learn Python the Hard Way (learnpythonthehardway.org)
by Zed Shaw
A great book and online resource; at the very least, read the introduction "The Hard Way Is Easier."

Python for Kids
by Jason R. Briggs
Again, more of a general programming book that happens to use Python (and written for younger readers).

5/Arduino and the Pi

As you'll see in the next few chapters, you can use the GPIO pins on the Raspberry Pi to connect to sensors or things like blinking LEDs and motors. And if you have experience using the Arduino microcontroller development platform, you can also use that alongside the Raspberry Pi.

When the Raspberry Pi was first announced, a lot of people asked if it was an Arduino killer. For about the same price, the Pi provides much more processing power, so why use an Arduino when you have a Pi? It turns out the two platforms are actually complementary, and the Raspberry Pi makes a great host for the Arduino. There are quite a few situations where you might want to put the Arduino and Pi together:

- To use the large number of libraries and sharable examples for the Arduino.
- To supplement an Arduino project with more processing power. For example, maybe you have a MIDI controller that was hooked up to a synthesizer, but now you want to upgrade to synthesizing the sound directly on the Pi.

- When you're dealing with 5V logic levels. The Pi operates at 3.3V, and its pins are not tolerant of 5V. The Arduino can act as a "translator" between the two.
- To prototype something a little out of your comfort zone, in which you may make some chip-damaging mistakes. For example, we've seen students try to drive motors directly from a pin on the circuit board (don't try it). On an Arduino, it was easy to pry the damaged microcontrollerchip out of its socket and replace it (less than $10 usually). Not so with the Raspberry Pi.
- When you have a problem that requires exact control in real-time, such as a controller for a 3D printer. As we saw in Chapter 4, the Raspberry Pi OS is not a real-time operating system, and programs can't depend on the same "instruction per clock cycles" rigor of a microcontroller.

The examples in this section assume that you know at least the basics of using the Arduino development board and integrated development environment (IDE). If you don't have a good grasp of the fundamentals, *Getting Started with Arduino* by Massimo Banzi and Michael Shiloh is a great place to start. The official Arduino tutorials (bit.ly/1oTWBNB) are quite good as well, and provide a lot of opportunities to cut and paste good working code.

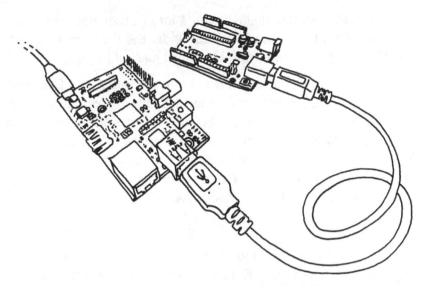

Figure 5-1. *Arduino and the Raspberry Pi are BFFs.*

Installing Arduino on the Raspberry Pi OS

To program an Arduino development board, you need to hook it up to a computer with a USB cable, then compile and flash a program to the board using the Arduino IDE. You can do this with any computer, or you can use your Raspberry Pi as a host to program the Arduino.

Using the Raspberry Pi to program the Arduino will be quicker to debug, and though compiling might be a *teensy* bit slower on the Pi, it's not going to be particularly noticeable. The Arduino IDE also only compiles code that has changed since the last compilation, so after the first compile, it's smooth sailing.

To install the Arduino IDE on the Raspberry Pi, type the following into a terminal:

```
sudo apt-get update      ❶
sudo apt-get install arduino      ❷
```

❶ Make sure you have the latest package list.
❷ Download the Arduino package.

This command will install Java plus a lot of other dependencies. The Arduino environment will appear under the *Programming* section of the program menu (don't launch it just yet though).

You can just plug the Arduino into one of the Raspberry Pi's open USB ports. The USB connection will be able to provide enough electrical power for the Arduino, but you might want to power the Arduino separately, depending on your application (if you're running motors or heaters, for instance). Too much of a power draw on the Pi's USB ports can cause it to act strangely, and can even cause a system crash and/or spontaneous reboot.

 Note that you'll need to plug the Arduino USB cable in after the Raspberry Pi has booted up. If you leave it plugged in at boot time, the Raspberry Pi may hang as it tries to figure out all the devices on the USB bus.

💡 When you launch the Arduino IDE, it polls all the USB devices and builds a list that is shown in the Tools→ Serial Port menu. Click Tools→Serial Port and select the serial port (most likely */dev/ttyACM0*), then click Tools→Board, and select the type of Arduino Board you have (e.g.,*Uno*). Click File→Examples→01. Basics→Blink to load a basic example sketch. Click the Upload button in the toolbar or choose File→Upload to upload the sketch, and after the sketch loads, the Arduino's onboard LED will start blinking.

To access the serial port on versions of the Raspberry Pi OS older than Jessie, you'll need to make sure that the *pi* user has permission to do so. You don't have to do this step on Raspbian Jessie. You can do that by adding the *pi* user to the *tty* and *dialout* groups. You'll need to do this before running the Arduino IDE:

```
sudo usermod ❶ -a -G  tty pi
sudo usermod -a -G dialout pi  ❷
```

❶   usermod is a Linux program to manage users.

❷   -a -G puts the user (pi) in the specified group (tty, then dialout).

Finding the Serial Port

If for some reason, */dev/ttyACM0* doesn't work, you'll need to do a little detective work. To find the USB serial port that the Arduino is plugged into without looking at the menu, try the following from the command line. Without the Arduino connected, type:

```
ls /dev/tty*
```

Plug in the Arduino, then try the same command again and see what changed. When I plugged in the Arduino, `/dev/ttyACM0` popped up in the listing.

Improving the User Experience

While you're getting set up, you may notice that the quality of the default font in the Arduino editor is less than ideal. You can improve it by downloading the open-source font Inconsolata. To install (when the Arduino IDE is closed), type:

```
sudo apt-get install fonts-inconsolata
```

Then edit the Arduino preferences file:

```
nano ~/.arduino/preferences.txt
```

and change the following lines to:

```
editor.font=Inconsolata,medium,14
editor.antialias=true
```

When you restart the Arduino IDE, the editor will use the new font.

Talking in Serial

To communicate between the Raspberry Pi and the Arduino over a serial connection, you'll use the built-in *Serial* library on the Arduino side, and the Python pySerial (github.com/ pyserial/pyserial) module on the Raspberry Pi side. It comes pre-installed in Jessie and subsequent releases, but if you ever need to install it, the package names are:

```
sudo apt-get install python-serial python3-serial
```

Open the Arduino IDE and upload this code to the Arduino:

```
void setup() {
    Serial.begin(9600);
}

void loop() {
    for(byte n = 0; n < 255; n++){
    Serial.write(n);
    delay(50);
    }
}
```

This counts upward and sends each number over the serial connection.

 Note that in Arduino, `Serial.write()` sends the actual number, which will get translated on the other side as an ASCII character code. If you want to send the string "123" instead of the number 123, use the `Serial.print()` command.

Next, you'll need to write a small Python script that will read the USB serial port the Arduino is connected to (see "Finding the Serial Port" on page 81). Here's the Python script; if the port isn't */dev/ttyACM0*, change the value of **port**. (See Chapter 4 for more on Python). Save it as *SerialEcho.py* and run it with **python SerialEcho.py**:

```
import serial

port = "/dev/ttyACM0"
serialFromArduino = serial.Serial(port,9600)   ❶
serialFromArduino.flushInput()   ❷
while True:
    if serialFromArduino.inWaiting() > 0:
        input = serialFromArduino.read(1)   ❸
        print(ord(input))   ❹
```

❶ Open the serial port connected to the Arduino.
❷ Clear out the input buffer.
❸ Read one byte from the serial buffer.
❹ Change the incoming byte into an actual number with **ord()**.

You won't be able to upload to the Arduino when Python has the serial port open, so make sure you kill the Python program with Ctrl-C before you upload the sketch again. You will be able to upload to an Arduino Micro, but doing so will break the connection with the Python script, so you'll need to restart it anyhow.

The Arduino is sending a number to the Python script, which interprets that number as a string. The `input` variable will contain whatever character maps to that number in the ASCII table (bit. ly/ZS47D0). To get a better idea, try replacing the last line of the Python script with this:

```
print(str(ord(input)), " = the ASCII character ", input, ".")
```

Setting the Serial Port as an Argument

If you want to set the port as a command-line argument in the Python sketch, use the `sys` module to grab the first argument:

```
import serial, sys

if (len(sys.argv) != 2):
    print("Usage:pythonReadSerial.pyport")
    sys.exit()
port = sys.argv[1]
```

After you do this, you can run the program like this:

```
python SerialEcho.py /dev/ttyACM0
```

The first simple example just sent a single byte; this could be fine if you are only sending a series of event codes from the Arduino. For example, if you have two buttons connected and the left button is pushed, send a 1; if the right, send 2. That's only good for 255 discrete events, though; more often you'll want to send arbitrarily large numbers or strings. If you're reading analog sensors with the Arduino, for example, you'll want to send numbers in the range of 0 to 1,023.

Parsing arbitrary numbers that come in one byte at a time is trivial, the way Python and pySerial handle strings. As a simple example, update your Arduino with the following code that counts from 0 to 1,024:

```
void setup() {
  Serial.begin(9600);
}

void loop() {
  for(int n = 0; n < 1024; n++)
    Serial.println(n, DEC);
    delay(50);
  }
}
```

The key difference is in the `println()` command. In the previous example, the `Serial.write()` function was used to write the raw number to the serial port. With `println()`, the Arduino formats the number as a decimal string and sends the ASCII codes for the string. So instead of sending a single byte with the value 254, it sends the string 254\r\n. The \r represents a carriage return, and the \n represents a new line (these are concepts that carried over from the teletypewriter into computing: carriage return moves to the start of the line; new line starts a new line of text).

On the Python side, you can use `readline()` instead of `read()`, which will read all of the characters up until (and including) the carriage return and new line. Python has a flexible set of functions for converting between the various data types and strings. It turns out you can just use the `int()` function to change the formatted string into an integer:

```
import serial

port = "/dev/ttyACM0"
serialFromArduino = serial.Serial(port,9600)
serialFromArduino.flushInput()
while True:
    input = serialFromArduino.readline()
    inputAsInteger = int(input)
    print(inputAsInteger * 10)
```

Note that it is simple to adapt this example so that it will read analog input and send the result; just change the loop to:

```
void setup() {
  Serial.begin(9600);
}

void loop(){
  int n = analogRead(A0);
  Serial.println(n, DEC);
  delay(100);
}
```

Assuming you change the Python script to just print `inputAsInteger` instead of `inputAsInteger * 10`, you should get some floating values in the 200 range if nothing is connected to analog pin 0. With some jumper wire, connect the Arduino's pin 0 to GND and the value should be 0. Connect it to the 3V3 pin and you'll see a value around 715, and 1,023 when connected to the 5V pin.

Using Arduino Compatibles

Many microcontroller boards are compatible with the Arduino IDE. Some use a special adapter from FTDI that handles all of the USB-to-TTL serial communication. To connect to these (often more inexpensive) boards, you would use an FTDI cable or an adapter board like the USB BUB or FTDI Friend. In the Jessie and subsequent releases, the FTDI driver for these boards is already installed, so they should work right out of the box. These boards typically show up as the */dev/ttyUSB0* device.

Using Firmata

As you go deeper, many projects will look the same as far as the code for basic communication goes. As with any form of communication, things get tricky once you get past "Hello, world"; you'll need to create *protocols* (or find an existing protocol and implement it) so that each side understands the other.

One good solution that comes bundled with the Arduino is Hans-Christoph Steiner's Firmata (bit.ly/1CXFnqF), an all-purpose serial protocol that is simple and human-readable. It may not be perfect for all applications, but it is a good place to start. Here's a quick example:

1. Select File→Examples→Firmata→StandardFirmata in the Arduino IDE; this will open the sample Firmata code that will allow you to send and receive messages to the Arduino and get information about all of the pins.

2. Upload that code to the Arduino the same way you did in previous examples.

3. You'll need a bit of Python code on the Pi to send commands to the Arduino to query or change its state. The easiest way is to use the pyFirmata module. Install it using Pip (see "EasyModule Installs with Pip" on page 71):

```
pip install --user pyfirmata
pip3 install --user pyfirmata
```

4. Because Firmata is a serial protocol, you talk to the Arduino from Python in the same way as in previous examples, but using pyFirmata instead of pySerial. Use the write() method to make a digital pin high or low on the Arduino:

```
from pyfirmata import Arduino
from time import sleep
board = Arduino('/dev/ttyACM0')
while (1):
    board.digital[13].write(1)
    print("on")
    sleep(1)
    board.digital[13].write(0)
    print("off")
    sleep(1)
```

This code makes the Raspberry Pi *blink the LED on the Arduino Uno board!* The full module is documented on the pyFirmata GitHub page (github.com/tino/pyFirmata).

Going Further

The nitty-gritty of serial protocols is beyond the scope of this book, but there are a lot of interesting examples of how other people have solved problems in the "Interfacing with Software" (bit.ly/1o17nGY) section of the Arduino Playground (www.arduino.cc/playground). In addition, you may want to try:

MIDI

If your project is musical, consider using MIDI commands as your serial protocol. MIDI is (basically) just serial, so it should just work.

Arduino-compatible Raspberry Pi shields

There are tons of shields, or PHATs (Pi HATs—Hardware Attached on Top) on the market that connect the GPIO pins on the Raspberry Pi with an Arduino-compatible microcontroller. WyoLum's AlaMode (bit.ly/1EylgRM) shield is a good solution and offers a few other accessories, including a real-time clock.

Talk over a network

Finally, you can ditch the serial connection altogether and talk to the Arduino over a network. A lot of really interesting projects are using the WebSocket (www.websocket.org) protocol along with the Node.js (nodejs.org) JavaScript platform.

Using the serial pins on the Raspberry Pi header

The header on the Raspberry Pi pulls out a number of input and output pins, including two that can be used to send and receive serial data bypassing the USB port. To do that, you'll first need to cover the material in Chapter 7, and make sure that you have a level shifter to protect the Raspberry Pi 3.3V pins from the Arduino's 5Vpins.

If you're looking to get deeper into making physical devices communicate, a good starting point is *Making Things Talk, 3rd Edition* (www.makershed.com/products/making-things-talk-third-edition), by Tom Igoe.

6/Basic Input and Output

While the Raspberry Pi is, in essence, a very inexpensive Linux computer, there are a few things that distinguish it from laptop and desktop machines that we usually use for writing email, browsing the Web, or word processing. One of the main differences is that the Raspberry Pi can be directly used in electronics projects, because it has *general-purpose input/output* pins right on the board, shown in Figure 6-1.

Figure 6-1. *Raspberry Pi's GPIO pins*

These GPIO pins can be accessed for controlling hardware such as LEDs, motors, and relays, which are all examples of outputs. As for inputs, your Raspberry Pi can read the status of buttons, switches, and dials, or it can read sensors for things like temperature, light, motion, or proximity (among many others).

 With the introduction of Raspberry Pi 1 Model B+, the number of GPIO pins increased from 26 to 40. If you have one of the first Pis, you can still carry out the examples in this chapter as they'll use only the first 26 pins on the GPIO header.

The best part of having a computer with GPIO pins is that you can create programs to read the inputs and control the outputs based on many different conditions, as easily as you'd program your desktop computer. Unlike a typical microcontroller board, which also has programmable GPIO pins, the Raspberry Pi has a few extra inputs and outputs, such as your keyboard, mouse, and monitor, as well as the Ethernet port, which can act as both an input and an output. If you have experience creating electronics projects with microcontroller boards like the Arduino, you have a few more inputs and outputs at your disposal with the Raspberry Pi. Best of all, they're built right in; there's no need to wire up any extra circuitry to use them.

Having a keyboard, mouse, and monitor is not the only advantage that Raspberry Pi has over typical microcontroller boards. There are a few other key features that will help you in your electronics projects:

Filesystem

Being able to read and write data in the Linux filesystem will make many projects much easier. For instance, you can connect a temperature sensor to the Raspberry Pi and have it take a reading once a second. Each reading can be appended to the end of a log file, which can be easily downloaded and parsed in a graphing program. It can even be graphed right on the Raspberry Pi itself!

Linux tools

Packaged in the Raspberry Pi's Linux distribution is a set of core command-line utilities, which let you work with files, control processes, and automate many different tasks. These powerful tools are at your disposal for all of your projects. And because there is an enormous community of Linux users that depend on these core utilities, getting help is usually one web search away. For general Linux help, you can usually find answers at Stack Overflow (stackoverflow.com). If you have a question specific to Raspberry Pi, try the Raspberry Pi Forum (www.raspberrypi.org/phpBB3) or the Raspberry Pi section of Stack Overflow (raspberrypi.stackexchange.com).

Languages

There are many programming languages out there, and embedded Linux systems like the Raspberry Pi give you the flexibility to choose whichever language you're most comfortable with. The examples in this book use shell scripting and Python, but they could easily be translated to languages like C, Java, or Perl.

One of the drawbacks to the Raspberry Pi is that there's no way to directly connect *analog sensors*, such as light and temperature sensors. Doing so requires a chip called an *analog-to-digital converter* or *ADC*. See Chapter 8 for how to read analog sensors using an ADC.

Using Inputs and Outputs

There are a few supplies that you'll need in addition to the Raspberry Pi itself to try out these basic input and output tutorials. Many of these parts you'll be able to find in hobby electronics component stores, or they can be ordered online from stores like Maker Shed, SparkFun, Adafruit, Mouser, or Digi-Key. Here are a few of the basic parts:

- Solderless breadboard
- LEDs, assorted
- Male-to-male jumper wires
- Female-to-male jumper wires (these are not as common as their male-to-male counterparts but are needed to connect the Raspberry Pi's GPIO pins to the breadboard)
- Push button switch
- Resistors, assorted

To make it easier to connect breadboarded components to the Raspberry Pi's pins, we also recommend Adafruit's Pi Cobbler Plus (adafruit.com/product/2029), which connects all the GPIO pins to a breadboard with a single ribbon cable. This eliminates the need to use female-to-male jumper wires.

In Figure 6-2, we've labeled each pin according to its default GPIO signal number, which is how you'll refer to a particular pin in the commands you execute and in the code that you write. The unlabeled pins are assigned to other functions by default.

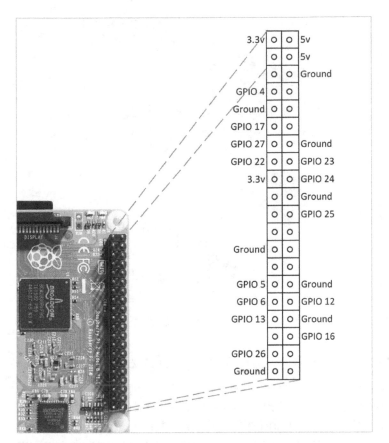

3.3v	o	o	5v
	o	o	5v
	o	o	Ground
GPIO 4	o	o	
Ground	o	o	
GPIO 17	o	o	
GPIO 27	o	o	Ground
GPIO 22	o	o	GPIO 23
3.3v	o	o	GPIO 24
	o	o	Ground
	o	o	GPIO 25
	o	o	
Ground	o	o	
	o	o	
GPIO 5	o	o	Ground
GPIO 6	o	o	GPIO 12
GPIO 13	o	o	Ground
	o	o	GPIO 16
GPIO 26	o	o	
Ground	o	o	

Figure 6-2. *The default GPIO pins on the Raspberry Pi. Some of the pins left blank could also be used as GPIO, but they have other possible functions. Unless you need more GPIO pins than are listed here, steer clear of them for now.*

There's a handy website created by Phil Howard called Raspberry Pinout (pinout.xyz) which we recommend you bookmark. It'll show you the Raspberry Pi's GPIO pins and has tons of reference information about how they can be used.

There are also great products such as RasPiO Portsplus port ID board (rasp.io), shown in Figure 6-3. It's a small board that fits over the GPIO pins for the sole purpose of making the pins easy to identify.

Figure 6-3. *The RasPiO Portsplus port ID board*

Digital Output: Lighting Up an LED

The easiest way to use outputs with the GPIO pins is by connecting a light emitting diode, or LED. You can then use the Linux command line to turn the LED on and off. Once you have an understanding of how these commands work, you're one step closer to having an LED light up to indicate when you have new email, when you need to take an umbrella with you as you leave your house, or when it's time to go to bed. It's also very easy to go beyond a basic LED and use a relay to control a lamp on a set schedule, for instance.

Beginner's Guide to Breadboarding

If you've never used a breadboard (Figure 6-4) before, it's important to know which terminals are connected. In the diagram, we've shaded the terminal connections on a typical breadboard. Note that the power buses on the left side are not connected to the power buses on the right side of the breadboard. You'll have to use male-to-male jumper cables to connect them to each other if you need ground and voltage on both sides of the breadboard.

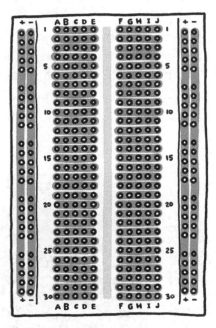

Figure 6-4. *Breadboard*

Here are the instructions you should follow:

1. Using a male-to-female jumper wire, connect pin 25 on the Raspberry Pi to the breadboard. Refer to Figure 6-2 for the location of each pin on the Raspberry Pi's GPIO header.

2. Using another jumper wire, connect one of the Raspberry Pi's ground pins to the negative power bus on the breadboard.

3. Now you're ready to connect the LED (see Figure 6-5). Before you do that, it's important to know that LEDs are *polarized*: it matters which of the LED's wires is connected to what. Of the two leads coming off the LED, the longer one is the anode (or "plus") and should be connected to a GPIO pin. The shorter lead is the cathode (or "minus") and should be connected to ground. Another way to tell the difference is by looking from the top. The flat side of the LED indicates the cathode, the side that should be connected to ground. Insert the anode side of the LED into the breadboard in the same channel as the jumper wire from pin 25, which will connect pin 25 to the LED. Insert the cathode side of the LED into the ground powerbus.

An easy way to remember the polarity of an LED is that the "plus" lead has had length *added* to it, while the "minus" lead has had length *subtracted* from it.

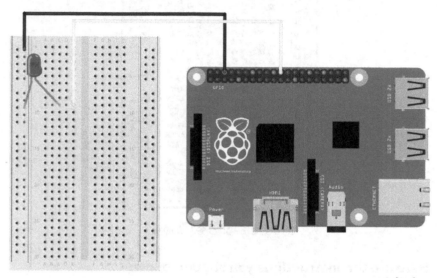

fritzing

Figure 6-5. *Connecting an LED to the Raspberry Pi*

4. With your keyboard, mouse, and monitor hooked up, power on your Raspberry Pi and log in. If you're at a command line, you're ready to go. If you're in the X Window environment, double-click the LXTerminal icon on your taskbar. This will bring up a terminal window.

5. To access the input and output pins from the command line, you'll need to run the commands as root, the super user account on the Raspberry Pi. To start running commands as root, type **sudo su** at the command line and press Enter:

```
pi@raspberrypi: ~ $ sudo su
root@raspberrypi:/home/pi#
```

You'll notice that the command prompt has changed from $ to #, indicating that you're now running commands as root.

The root account has administrative access to all the functions and files on the system and there is very little protecting you from damaging the operating system if you type a command that can harm it. You must exercise caution when running commands as root. If you do mess something up, don't worry about it too much; you can always reimage the SD card with a clean Linux install; however, you'll lose any customization you made to the operating system, as well as any programs or sketches you wrote.

When you're done working within the root account, type **exit** to return to working within the **pi** user account.

---

6. Before you can use the command line to turn the LED on pin 25 on and off, you need to export the pin to the userspace (in other words, make the pin available for use outside of the confines of the Linux kernel), this way:

```
root@raspberrypi:/home/pi# echo 25 > /sys/class/gpio/export
```

The echo command writes the number of the pin you want to use (25) to the export file, which is located in the folder */sys/class/gpio*. When you write pin numbers to this special file, it creates a new directory in */sys/class/gpio* that has the control files for the pin. In this case, it created a new directory called */sys/class/gpio/gpio25*.

7. Change to that directory with the **cd** command and list the contents of it with **ls**:

```
root@raspberrypi:/home/pi#cd/sys/class/gpio/gpio25
root@raspberrypi:/sys/class/gpio/gpio25# ls
active_low direction edge power subsystem uevent
   value
```

The command **cd** stands for "change directory." It changes the working directory so that you don't have to type the full path for every file. **ls** will list the files and folders within that directory. There are two files that you're going to work within this directory: direction and value.

8. The direction file is how you'll set this pin to be an input (like a button) or an output (like an LED). Because you have an LED connected to pin 25 and you want to control it, you're going to set this pin as an output:

```
root@raspberrypi:/sys/class/gpio/gpio25# echo out >
direction
```

9. To turn the LED on, you'll use the echo command again to write the number 1 to the value file:

```
root@raspberrypi:/sys/class/gpio/gpio25# echo 1 >
value
```

10. After pressing Enter, the LED will turn on! Turning it off is as simple as using echo to write a zero to the value file:

```
root@raspberrypi:/sys/class/gpio/gpio25# echo 0 > value
```

Virtual Files

The files that you're working with aren't actually files on the Raspberry Pi's SD card, but rather are a part of Linux's *virtual filesystem*, which is a system that makes it easier to access low-level functions of the board more simply. For example, you could turn the LED on and off by writing to a particular section of the Raspberry Pi's memory, but doing so would require more coding and more caution.

So if writing to a file is how you control components that are outputs, how do you check the status of components that are inputs? If you guessed "reading a file," then you're absolutely right. Let's try that now.

Digital Input: Reading a Button

Simple push button switches like the one in Figure 6-6 are great for controlling basic digital input. Best of all, they're made to fit perfectly into a breadboard.

These small buttons are very commonly used in electronics projects and understanding what's going on inside of them will help you as you prototype your project. When looking at the button as it sits on the breadboard (see Figure 6-6), the top two terminals are always connected to each other. The same is true for the bottom two terminals; they're always connected. When you push down on the button, these two sets of terminals are connected.

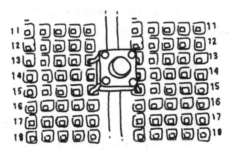

Figure6-6. *Button*

When you read a digital input on a Raspberry Pi, you're check-ing to see if the pin is connected to either 3.3V or to ground. It's important to remember that it must be either one or the other, and if you try to read a pin that's not connected to either 3.3V or ground, you'll get unexpected results. Once you understand how digital input with a push button works, you can start using com-ponents like magnetic security switches, arcade joysticks, or even vending machine coin slots. Start by wiring up a switch to read its state:

1. Insert the push button into the breadboard so that its leads straddle the middle channel.

2. Using a jumper wire, connect pin 24 from the Raspberry Pi to one of the top terminals of the button.

3. Connect the 3V3 pin from the Raspberry Pi to the positive powerbus on the breadboard.

Be sure that you connect the button to the 3V3 pin and not the 5V pin. Using more than 3.3V on an input pin will permanently damage your Raspberry Pi.

4. Connect one of the bottom terminals of the button to the power bus. Now when you push down on the button, the 3.3V will be connected to pin 24.

5. Remember what we said about how a digital input must be connected to *either* 3.3V or ground? When you let go of the button, pin 24 isn't connected to either of those and is therefore *floating*. This condition will cause unexpected results, so let's fix that. Use a 10K resistor (labeled with the colored bands: brown, black, orange, and then silver or gold) to connect the input side of the switch to the ground rail, which you connected to the Raspberry Pi's ground in the output example. When the switch is not pressed, the pin will be connected to ground. Electricity always follows the path of least resistance toward ground, so when you press the switch, the 3.3V will go toward the Raspberry Pi's input pin, which has less resistance than the 10K resistor. When everything's hooked up, it should look like Figure 6-7.

6. Now that the circuit is built, let's read the value of the pin from the command line. If you're not already running commands as root, type **sudo su**.

7. As with the previous example, you need to export the input pin to userspace:

```
# echo 24 > /sys/class/gpio/export
```

8. Let's change to the directory that was created during the export operation:

```
# cd /sys/class/gpio/gpio24
```

9. Now set the direction of the pin to input:

```
# echo in > direction
```

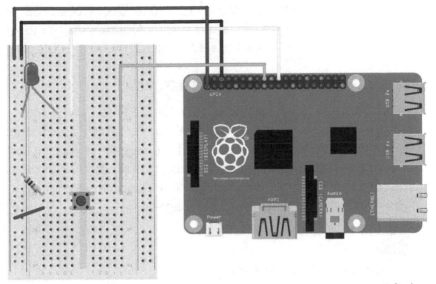

Figure 6-7. *Connecting a button to the Raspberry Pi*

10. To read the value of the pin, you'll use the cat command, which will print the contents of files to the terminal. The command cat gets its name because it can also be used to concatenate, or join, files. It can also display the contents of a file for you:

```
# cat value
0
```

11. The zero indicates that the pin is connected to ground. Now press and hold the button while you execute the command again:

```
# cat value
1
```

12. If you see the number 1, you'll know you've got it right!

To easily execute a command that you've previously executed, hit the up arrow key until you see the command that you want to run and then hit Enter.

Now that you can use the Linux command line to control an LED or read the status of a button, let's use a few of Linux's built-in tools to create a very simple project that uses digital input and output.

Project: Cron Lamp Timer

Let's say you're leaving for a long vacation early tomorrow morning and you want to ward off would-be burglars from your home. A lamp timer is a good deterrent, but hardware stores are closed for the night and you won't have time to get one before your flight in the morning. However, because you're a Raspberry Pi hobbyist, you have a few supplies lying around, namely:

* Raspberry Pi board
* Breadboard
* Jumper wires, female-to-male
* PowerSwitch Tail II relay
* Hookup wire

With these supplies, you can make your own programmable lamp timer using two powerful Linux tools: *shell scripts* and *cron*.

Scripting Commands

A shell script is a file that contains a series of commands (just like the ones you've been using to control and read the pins). Take a look at the following shell script and the explanation of the key lines:

```
#!/bin/bash  ❶
echo Exporting pin $1.  ❷
echo $1 > /sys/class/gpio/export
echo Setting direction of $1 to out.  ❸
echo out > /sys/class/gpio/gpio$1/direction  ❹
echo Setting pin $1 high.
echo 1 > /sys/class/gpio/gpio$1/value
```

❶ This line is required for all shell scripts.

❷ $1 refers to the first command-line argument.

❸ Instead of exporting a specific pin number, the script uses the first command-line argument.

❹ Notice that the first command-line argument replaces the pin number here as well.

Save that as a text file called *on.sh* and make it executable with the chmod command:

```
root@raspberrypi:/home/pi # chmod +x on.sh
```

 You still need to be executing these commands as root. Type sudo su if you're getting errors like "Permission denied."

A command-line argument is a way of passing information into a program or script by typing it in after the name of the command. When you're writing a shell script, $1 refers to the first command-line argument, $2 refers to the second, and so on. In the case of on.sh, you'll type in the pin number that you want to export and turn on. Instead of *hard coding* pin 25 into the shell script, it's more universal by referring to the pin that was typed in at the command line. To export pin 25 and turn it on, you can now type:

```
root@raspberrypi:/home/pi/# ./on.sh 25   ❶
Exporting pin 25.
Setting direction of 25 to out.
Setting pin 25 high.
```

❶ The ./ before the filename indicates that you're executing the script in the directory you're in.

If you still have the LED connected to pin 25 from earlier in the chapter, it should turn on. Let's make another shell script called *off. sh*, which will turn the LED off. It will look like this:

```
#!/bin/bash
echo Setting pin $1 low.
echo 0 >/sys/class/gpio/gpio$1/value
echo Unexporting pin $1
echo $1 > /sys/class/gpio/unexport
```

Now let's make it executable and run the script:

```
root@raspberrypi:/home/pi/temp# chmod +x off.sh
root@raspberrypi:/home/pi/temp# ./off.sh25
Settingpin25low.
Unexportingpin25
```

If everything worked, the LED should have turned off.

Connecting a Lamp

Of course, a tiny little LED isn't going to give off enough light to fool burglars into thinking that you're home, so let's hook up a lamp to the Raspberry Pi:

1. Remove the LED connected to pin 25.

2. Connect two strands of hookup wire to the breadboard, one that connects to pin 25 of the Raspberry Pi and the other to the ground bus.

3. The strand of wire that connects to pin 25 should be connected to the "+in" terminal of the PowerSwitch Tail.

4. The strand of wire that connects to ground should be connected to the "-in" terminal of the PowerSwitch Tail. Compare your circuit to Figure 6-8.

5. Plug the PowerSwitch Tail into the wall and plug a lamp into the PowerSwitch Tail. Be sure the lamp's switch is in the on position.

6. Now when you execute `./on.sh 25`, the lamp should turn on. If you execute `./off.sh 25`, the lamp should turn off!

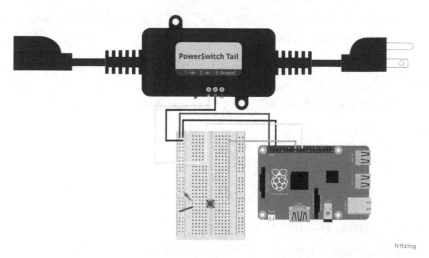

Figure 6-8. *Connecting a PowerSwitch Tail II to the Raspberry Pi*

Inside the PowerSwitch Tail, there are a few electronic components that help you control high-voltage devices like a lamp or blender by using a low-voltage signal such as the one from the Raspberry Pi. The "click" you hear from the PowerSwitch Tail when it's turned on or off is the relay, the core component of the circuit inside. A relay acts like a switch for the high-voltage device that can be turned on or off depending on whether the low-voltage control signal from the Raspberry Pi is on or off.

Scheduling Commands with cron

So now you've packaged up a few different commands into two-simple commands that can turn a pin on or off. And with the lamp connected to the Raspberry Pi through the PowerSwitch Tail, you can turn the lamp on or off with a single command. Now you can use cron to schedule the light to turn on and off at different times of the day. cron is Linux's job scheduler. With it, you can set commands to execute on specific times and dates, or you can have jobs run on a particular period (for example, once an hour). You're going to schedule two jobs: one of them will turn the light on at 8:00 p.m., and the other will turn the light off at 2:00 a.m.

As with other time-dependent programs, you'll want to make sure you've got the correct date and time set up on your Raspberry Pi, as described in "Setting the Date and Time" on page 47.

To add these jobs, you'll have to edit the cron table (a list of commands that Linux executes at specified times):

```
root@raspberrypi:/home/pi/# crontab -e
```

This will launch a text editor to change root's cron table (to change to the root user, type sudo su). At the top of the file, you'll see some information about how to modify the cron table. Use your arrow keys to get to the bottom of the file and add these two entries at the end of the file:

```
0 20 * * * /home/pi/on.sh 25
0 2 * * * /home/pi/off.sh 25
```

---

 cron will ignore any lines that start with the hash mark. If you want to temporarily disable a line without deleting it or add a comment to the file, put a hash mark in front of the line. cron also expects a blank line at the end of the file; if you get unexpected behavior from your cron table, make sure you haven't filled that blank line with text/commands and not replaced it.

Press Ctrl-X to exit, press y to save the file when it prompts you, and hit Enter to accept the default filename. When the file is saved and you're back at the command line, it should say installing new crontab to indicate that the changes you've made are going to be executed by cron.

More About cron

cron will let you schedule jobs for specific dates and times or at intervals. There are five time fields (or six if you want to schedule by year), each separated by a space followed by another space, then the command to execute; asterisks indicate that the job should execute each period (Table 6-1).

Table 6-1. *cron entry for turning light on at 8:00 p.m. everyday*

0	20	*	*	*	/home/pi/on.sh25
Minute (:00)	Hour (8 p.m.)	Everyday	Every month	Every day of the week	Path to command

Let's say you only wanted the lamp to turn on every weekday. Table 6-2 shows what the crontab entry would look like.

Table 6-2. *cron entry for turning light on at 8:00 p.m. every weekday*

0	20	*	*	1-5	/home/pi/on.sh25
Minute (:00)	Hour (8 p.m.)	Every-day	Every month	Monday to Friday	Path to command

As another example, let's say you have a shell script that checks if you have new mail and emails you if you do. Table 6-3 shows how you'd get that script to run every five minutes.

Table 6-3. *cron entry for checking for mail every five minutes*

*/5	*	*	*	*	/home/pi/checkMail.sh
Every five minutes	Every hour	Every-day	Every month	Every day of the week	Path to command

The */5 indicates a period of every five minutes.

As you can see, `cron` is a powerful tool that's at your disposal for scheduling jobs for specific dates or times and at specific intervals.

Going Further

eLinux's Raspberry Pi GPIO Reference Page
(bit.ly/1vORWl3)

This is the most comprehensive reference guide to the Raspberry Pi's GPIO pins.

Gordon Henderson's Command Line GPIO Utility
(projects.drogon.net/raspberry-pi/wiringpi/the-gpio-utility)

This command line utility makes it easier to work with GPIO pins from the command line. It's bundled with all the latest versions of the Raspberry Pi OS. Try running the command `gpioreadall` to get an overview of all your pins.

7/Programming Inputs and Outputs with Python

At the end of Chapter 6, you did a little bit of programming with the Raspberry Pi's GPIO pins using a shell script. In this chapter, you're going to learn how to use Python to do the same thing... and a little more. As with the shell script, Python will let you access the GPIO pins with code that reads and controls the pins automatically.

The advantage that Python has over shell scripting is that Python code is easier to write and is more readable. There's also a whole slew of Python modules that make it easy for you to do some complex stuff with basic code. See Table 4-2 for a list of a few modules that might be useful in your projects. Best of all, there's a Python module called GPIO Zero (gpiozero.readthedocs.io/en/stable) that makes it easy to read and control the GPIO pins. You're going to learn how to use that module in this chapter.

Installation

GPIO Zero is now included in the desktop version of the Raspberry PI OS. It builds on some other libraries that you may be familiar with, including RPi.GPIO (bit.ly/1vzTBtl) and pigpio, and even lets you select particular pin libraries as you need them.

If you are using Raspberry Pi OS Lite or another OS on the Pi, you'll need to install GPIO Zero, which is not a difficult task.

With the Pi OS Lite, first update your repositories list with

```
pi@raspberrypi: ~ $ sudo apt update
```

Then install either the Python 2 or Python 3 package:

```
pi@raspberrypi: ~ $ sudo apt install python-gpiozero
```

or

```
pi@raspberrypi: ~ $ sudo apt install python3-gpiozero
```

If you're using another OS, you may need to use pip to install it instead. Use

```
pi@raspberrypi: ~ $ sudo pip install gpiozero
```

or

```
pi@raspberrypi: ~ $ sudo pip3 install gpiozero
```

for Python 2 or 3, respectively.

 There are some programmers, myself included, who frown on using **sudo** when installing packages with **pip**. This is because using **sudo** may install some of the dependencies and binaries in directories that are inaccessible to ordinary, non-root users. That can often lead to problems when trying to use the packages that were installed this way. Most often I will use the **--user** flag, like so:

```
pipinstall --usergpiozero
```

Only if the installation fails that way will I use the **sudo** method.

Testing GPIO in Python

1. Go into the Python interactive interpreter from the terminal prompt by typing `python3`.

2. When you're at the >>> prompt, try importing the LED module:

```
>>> from gpio zero import LED
```

If you don't get any errors after entering the import command, you know it's installed and ready to use.

1. As with other libraries, gpiozero allows you to control each of the Pi's pins individually. However, there's a catch. gpiozero refers to the pins not by their physical placement on the board, but by their internal connection to the CPU, which is what's known as the pin's Broadcom or BCM number. Other libraries allow you to select how you're going to refer to the pins; gpiozero only gives you the Broadcom option. Luckily, it's pretty easy to find printouts online of the pins' individual BCM numbers. There is also an awesome mobile app called Electrodoc (play.google.com/store/apps/details?id=it.android.demi. elettronica&hl=en_US&gl=US or apps.apple.com/us/app/ electrodoc-pro/id1146647134) that I have installed on my phone, which lets you call up the pin-outs for almost any hobbyist board you can think of. I highly recommend it!

2. Tell the gpiozero library which pin to use. Let's use GPIO pin #4, and connect it the way you did in "Beginner's Guide to Breadboarding":

3. `>>> led = LED(4)`

4. Turn on the LED:
```
>>> led.on()
```

5. Turn off the LED:
```
>>> led.off()
```

6. Exit the Python interactive interpreter:
```
>>> exit()
$
```

In Chapter 6, you learned that digital input and output signals on the Raspberry Pi must be either 3.3V or ground. In digital electronics, we refer to these signals as high or low, respectively. Keep in mind that not all hardware out there uses 3.3V to indicate high; some use 1.8V, and others use 5V. If you plan on connecting your Raspberry Pi to digital hardware through its GPIO pins, it's important that the other hardware also uses 3.3V.

These steps give you a rough idea of how to control the GPIO pins by typing Python statements directly into the interactive interpreter. It's hard to imagine it being any easier, to be honest! Just as you created a shell script to turn the pins on and off in Chapter 6, you're going to create a Python script to read and control the pins automatically.

Blinking an LED

To blink an LED on and off with Python, you're going to use the statements that you already tried in the interactive interpreter, in addition to a few others. For the next few steps, we'll assume you're using the desktop environment (as shown in Figure 7-1), but feel free to use the command line to write and execute these Python scripts if you prefer.

Here's how to blink an LED with a Python script:

1. Open the text editor by clicking the Raspberry menu→ Accessories→TextEditor.

2. Enter the following code:

```
from gpizero import LED  ❶
import time  ❷

led = LED(4)  ❸

while True:  ❹
```

```
led.on()  ❺
time.sleep(1)  ❻
led.off()  ❼
time.sleep(1)  ❽
```

❶ Import the code needed for GPIO control.

❷ Import the code needed for the sleep function.

❸ Set pin 4 as an output for the LED.

❹ Create an infinite loop consisting of the indented code below it.

❺ Turn the LED on.

❻ Wait for one second.

❼ Turn the LED off.

❽ Wait for one second.

3. Save the file as *blink.py* within the home directory, */home/pi*. There's a shortcut to this folder in the places list on the left side of the Save As window.

 Remember—indentation matters in Python!

---

4. Open LXTerminal, then use these commands to make sure the working directory is your home directory, and execute the script (see Figure 7-1):

```
pi@raspberrypi ~/Documents $ cd ~
pi@raspberrypi ~ $ python blink.py
```

5. Your LED should now be blinking!

6. Press Ctrl-C to stop the script and return to the command line.

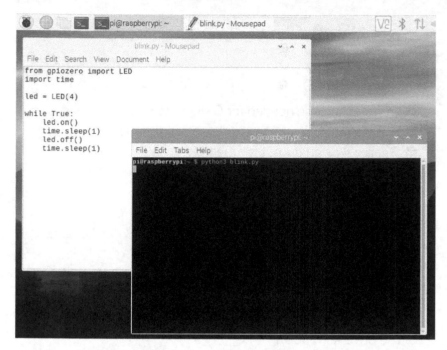

Figure 7-1. *Using a text editor and LXTerminal to edit and launch Python scripts.*

Try modifying the script to make the LED blink faster by using decimals in the `time.sleep()` functions. You can also try adding a few more LEDs and getting them to blink in a pattern. You can use any of the dedicated GPIO pins that are shown in Figure 6-2.

Reading a Button

If you want something to happen when you press a button, one way to do that is to use a technique called *polling*. Polling means continually checking over and over again for some condition. In this case, your program will be polling whether the button is connecting the input pin to 3.3V or to ground. To learn about polling, you'll create a new Python script that will display text on the screen when the user pushes a button:

1. Connect a push button between pin 24 and ground. (Remember, you're using BCM numbers to refer to the pins, so here we're actually referring to physical pin #18!) gpio-zero's Button.wait_for_press() function checks for a pin going to ground. The logic behind doing it this way is that you don't need to connect a pulldown resistor between the switch and GND.

2. Create a file in your home directory called *button.py* and open it in the editor.

3. Enter the following code:

```python
from gpiozero import Button
import time
button = Button(24)   ❶
count = 0 ❷
while True:
    button.wait_for_press()   ❸
    count += 1  ❹
    print("Buttonpressed" + str(count) + "times.")  ❺
    time.sleep(0.1)  ❻
```

❶ Set pin 24 as an input.
❷ Create a variable called **count** and store 0 in it.
❸ Poll the button for presses.
❹ Add to count if the button has been pressed.
❺ Print the text to the terminal
❻ Wait briefly, but let other programs have a chance to run by not hogging the processor's time.

4. Go back to LXTerminal and execute the script:

```
$ python3 button.py
```

5. Now press the button. If you've got everything right, you'll see a few lines of "The button has been pressed" for each time you press the button.

The preceding code checks for the status of the button 10 times per second, which is why you'll see more than one sentence printed (unless you have incredibly fast fingers). The Python statement `time.sleep(0.1)` is what controls how often the button is checked.

But why not continually check the button? If you were to remove the `time.sleep(0.1)` statement from the program, the loop would indeed run incredibly fast, so you'd know much more quickly when the button was pressed. This comes with a few drawbacks: you'd be using the processor on the board constantly, which will make it difficult for other programs to function, and it would increase the Raspberry Pi's power consumption. Because *button.py* has to share resources with other programs, you have to be careful that it doesn't hog them all up.

These are challenges that you'll face when you're using polling to check the status of a digital input. One way to get around these challenges is to use an *interrupt*, which is a way of setting a specified block of code to run when the hardware senses a change in the state of the pin. gpiozero supports interrupts—specifically, detecting and responding to an edge *state change* (when the pin changes from HIGH to LOW, or vice versa) and you can read about how to use that feature in the library's documentation (gpiozero. readthedocs.io/en/stable/api_pins.html).

Note that edge detection only works on pins that support it, and the library documents state that depending on the hardware, the information read from the state at any given time is not guaranteed to be accurate.

Project: Simple Soundboard

Now that you know how to read the inputs on the Raspberry Pi, you can use the sound functions of the Python module Pygame to make a soundboard. A soundboard lets you play small sound recordings when you push its buttons. To make your own soundboard, you'll need the following in addition to your Raspberry Pi:

- Three push button switches
- Female-to-male jumper wires
- Standard jumper wires or hookup wire, cut to size
- Solderless breadboard
- Computer speakers, or an HDMI monitor that has built-in speakers. You can also use a simple pair of headphones—anything that will plug into the Pi's audio analog OUT port.

You'll also need a few uncompressed sound files, in *.wav* format. For purposes of testing, there are a few sound files preloaded on the Raspberry Pi that you can use. Once you get the soundboard working, it's easy to replace those files with any sounds you want, though you may have to convert them to *.wav* from other formats. Start by building the circuit:

1. Using a female-to-male jumper wire, connect the Raspberry Pi's ground pin to the negative rail on the breadboard.

2. Insert the three push button switches in the breadboard, all straddling the center trench.

3. Using standard jumper wires or small pieces of hookup wire, connect the groundrail of the breadboard to the top pin of each button.

4. Using female-to-male jumper wires, connect each button's bottom pin to the Raspberry Pi's GPIO pins. For this project, we used pins 4, 14, and 25.

Figure 7-2 shows the completed circuit. We created this diagram with Fritzing (fritzing.org), an open-source tool for creating hardware designs.

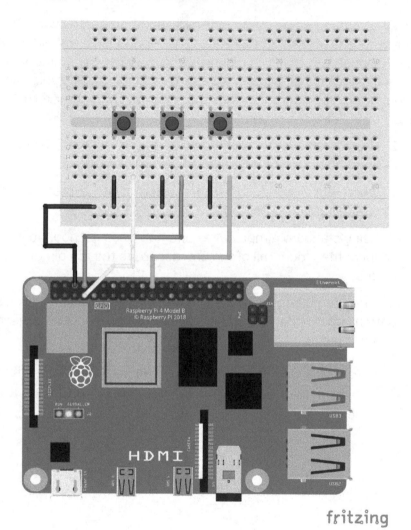

Figure 7-2. *Completed circuit for the soundboard project*

Now that you have the circuit breadboarded, it's time to work on the code:

1. Create a new directory in your home directory called *sound-board*.

2. Open that folder and create a file there called *soundboard.py*.

3. Open *soundboard.py* and type in the following code:

```
import pygame.mixer
from time import sleep
from gpiozero import Button
from sys import exit

button1 = Button(4)
button2 = Button(14)
button3 = Button(25) Again, BCM 4,14 and 25 match up
with pins 4,14, and 25??

pygame.mixer.init(48000, -16, 1, 1024)  ❶
soundA = pygame.mixer.Sound(  ❷
    "/usr/share/sounds/alsa/Front_Center.wav")
soundB = pygame.mixer.Sound(
    "/usr/share/sounds/alsa/Front_Left.wav")
soundC = pygame.mixer.Sound(
    "/usr/share/sounds/alsa/Front_Right.wav")

soundChannelA = pygame.mixer.Channel(1)  ❸
soundChannelB = pygame.mixer.Channel(2)
soundChannelC = pygame.mixer.Channel(3)

print "Soundboard Ready."  ❹

while True:
    try:
        button1.wait_for_press()  ❺
        soundChannelA.play(soundA)  ❻
        button2.wait_for_press()
        soundChannelB.play(soundB)
        button3.wait_for_press()
        soundChannelC.play(soundC)
        sleep(.01)  ❼

    except KeyboardInterrupt:  ❽
    exit()
```

❶ Initialize Pygame's mixer.

❷ Load the sounds.

❸ Setup three channels (one for each sound) so that we can play different sounds concurrently.

❹ Let the user know the soundboard is ready.

❺ Poll the button. If the button is pressed, execute the following line.

❻ Play the sound.

❼ Don't "peg" the processor by checking the buttons faster than we need to.

❽ This will let us exit the script cleanly when the user hits Ctrl-C, without showing the trace back message.

4. Go to the command line and navigate to the folder where you've saved *soundboard.py* and execute the script:

```
pi@raspberrypi ~/soundboard $ python3 soundboard.py
```

5. After you see "Soundboard Ready," start pushing the buttons to play the sound samples.

Depending on how your Raspberry Pi is setup, your sound might be sent via HDMI to your display, or it may be sent to the 3.5mm analog audio output jack on the board. To change that, exit out of the script by pressing Ctrl-C and executing the following command to use the analog audio output:

```
pi@raspberrypi ~/soundboard $ sudo amixer cset numid=31
```

To send the audio through HDMI to the monitor, use:

```
pi@raspberrypi ~/soundboard $ sudo amixer cset numid=32
```

Of course, the stock sounds aren't very interesting, but you can replace them with any of your own sounds: applause, laughter, buzzers, and dings. Add them to the *soundboard* directory and update the code to point to those files. If you want to use more sounds on your soundboard, add additional buttons and update the code as necessary.

Going Further

gpiozero documentation
(gpiozero.readthedocs.io/en/stable/)
The gpiozero library is not only under active development, it's got a ton of functionality that we've barely touched on here. It would be well worth your time to dig through the documentation, breadboard up some components, and see what you can do with it.

8/Analog Input
and Output

In Chapter 6, you learned about digital inputs and outputs with buttons, switches,
LEDs, and relays. Each of these components was always either on or off, never
anything in between. However, you might
want to sense things in the world that are
not necessarily one or the other—for instance, temperature, distance, light levels,
or the status of a dial. These all come in
a range of values.

Or you may want to put something in a state that's not just on or
off. For example, suppose you wanted to dim an LED or control the
speed of a motor, rather than just turning it on or off. To make an
analogy for analog and digital, you can think of a typical light switch
versus a dimmer switch, as pictured in Figure 8-1.

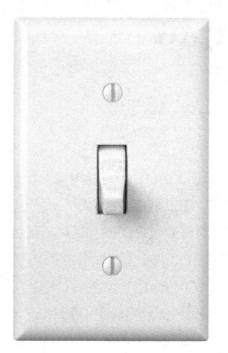
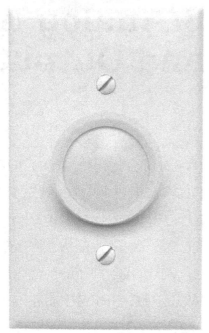

Figure 8-1. *Digital is like the switch on the left: it can be either on or off. Analog, on the other hand, can be set at a range of values between fully on and completely off.*

Output: Converting Digital to Analog

Just as in Chapter 7, in this chapter you'll use the GPIO Zero module already installed in the most recent versions of Raspberry Pi OS. The module has functions for controlling the GPIO pins. Some of these functions act sort of like a dimmer switch.

We say that it's "sort of" like a dimmer switch because the module uses a method called *pulse-width modulation*, or PWM, to make it *seem* like there's a range of voltages coming out of its outputs. What the GPIO module is actually doing is pulsing its pins on and off really quickly, so quickly that the human eye doesn't register the blinking. All you see is that the light isn't as bright as it would be if it were on all the time. If you want the pin to behave as though it's at half voltage, the pin will be pulsed so that it is on 50% of the time and off 50% of the time. If you want the pin to behave as though it's at 20% power, the Raspberry Pi will turn the pin on 20% of the time and off 80% of the time. The percentage of time that it's on versus the total time of the cycle is called the *duty cycle* (Figure 8-2). When you connect an LED to these pins and instruct the Raspberry Pi to change the duty cycle, the effect we see is that the LED gets dimmer.

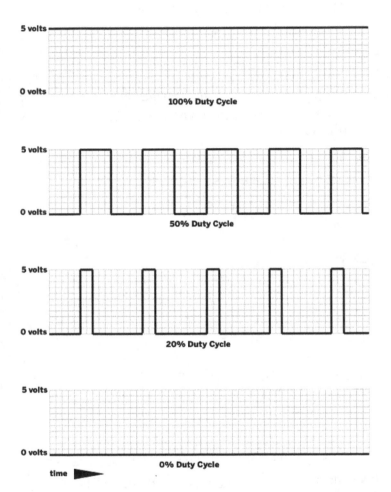

Figure 8-2. *The duty cycle represents how much time the pin is turned on over the course of an on-off cycle.*

Test-Driving PWM

For the next few steps, we'll assume you're using the desktop environment, but feel free to use the command line to write and execute these Python scripts if you prefer:

1. Connect an LED to GPIO pin 25 (physical pin 22) as you did in "Beginner's Guide to Breadboarding" on page 94.

2. Open the File Manager by clicking its icon in the taskbar.

3. Be sure you're in the home directory, the default being / *home/pi*. If not, click on the home icon under the Places listing.

4. Create a file in your home directory called *blink.py*. Do this by right-clicking in the home directory window and choosing "New File." Name the file *fade.py*.

5. Double-click on *fade.py* to open it in the default text editor.

6. Enter the following code and save the file:

```
from gpiozero import PWMLED   ❶
from time import sleep

led = PWMLED(25)

while True:
    for dc in range(0, 100, 1):   ❷
        led.value = dc/100.0   ❸
        time.sleep(0.01)
    for dc in range(100, 0, -1):   ❹
        led.value = dc/100.0
        time.sleep(0.01)
```

❶ Create a PWM LED object and set it to GPIO pin 25.

❷ Run the indented code below this line, each time incrementing the value of **dc** by 1 from starting at 0 and going to 100.

❸ Set the duty cycle of GPIO pin 25 to the value of **dc**.

❹ Run the indented code that follows, each time decrementing the value of **dc** by 1 from starting at 100 and going to 0.

7. Open the terminal, then use these commands to make sure the working directory is your home directory, and execute the script:

```
pi@raspberrypi:~/Development $ cd~
pi@raspberrypi:~ $ python3 fade.py
```

8. Your LED should now be fading up and down!

9. Press Ctrl-C to stop the script and return to the command line.

 If you're accustomed to using PWM on a microcontroller like the Arduino, you'll find that—unlike Arduino—there is unsteadiness in the PWM pulses from the Raspberry Pi. This is because in this example, you're using the CPU to turn the LED on and off. Because that CPU is used for multiple things at one time, it may not always keep perfect time. You can always reach for other hardware like Adafruit's PWM/Servo Driver (www.adafruit.com/products/815) if you need to have more precise control.

Taking PWM Further

With the ability to use pulse-width modulation to fade LEDs up and down, you could also connect an RGB LED and control the color by individually changing the brightness of its red, green, and blue elements.

As we mentioned earlier, you can also use pulse-width modulation to control the speed of a direct current motor that's connected to your Raspberry Pi through transistors. The PWM output, when fed into the transistors, will modulate the amount of power the transistors allow into the motor, and hence its speed.

The position of the shaft on a hobby servo motor (the kind that steers RC cars) can also be controlled with specific pulses of electricity. Though, keep in mind that you may need additional hardware and power to control these motors with a Raspberry Pi.

Input: Converting Analog to Digital

Just like you controlled the output of a GPIO pin on a scale of 0 to 100, it's also possible to read sensors that can offer the Raspberry Pi a range of values. If you want to know the temperature outdoors, the light level of a room, or the amount of pressure on a resistive pad, you can use various sensors. On a microcontroller like the Arduino, there's special hardware to convert the analog voltage level to digital data. Unfortunately, your Raspberry Pi doesn't have this hardware built-in.

This section will show you how to convert from analog to digital using an *ADC*, or *analog-to-digital converter*. There are a few different models of ADCs out there, but this chapter will use the ADS1x15 from Texas Instruments. The package of the ADS1x15 chip is too small for your standard breadboard, so Adafruit Industries has created a breakout board (www.adafruit.com/products/1083) for it, shown in Figure 8-3. Once you've soldered header pins to the breakout board, you can build prototypes with this chip in your breadboard. The chip uses a protocol named *Inter-Integrated Circuit* (commonly called *I²C*) for transmitting the analog readings. Luckily, we don't need to understand the protocol to use it. Adafruit provides an excellent open-source Python library to read the values from the ADS1115 or its little brother, the ADS1015, via I²C.

To connect the ADS1115 or ADS1015 breakout to your Raspberry Pi:

1. Connect the 3.3V pin from the Raspberry Pi to the positive rail of the breadboard. Refer to Figure 6-2 for pin locations on the Raspberry Pi's GPIO header

2. Connect the ground pin from the Raspberry Pi to the negative rail of the breadboard.

3. Insert the ADS1x15 into the breadboard and use jumper wires to connect its VDD pin to the positive rail and its GND pin to the negative rail.

4. Connect the SCL pin on the ADS1x15 to the SCL pin on the Raspberry Pi. The SCL pin on the Pi is the one paired with the ground pin on the GPIO header.

5. Connect the SDA pin on the ADS1x15 to the SDA pin on the Raspberry Pi. The SDA pin is in between the SCL pin and the 3.3V pin.

Figure 8-3. *The ADS1015 analog-to-digital converter breakout from Adafruit*

Now you'll need to connect an analog sensor to the ADS1x15. There are many to choose from, but for this walk-through, you'll use a 2K potentiometer so that we can have a dial input for your Raspberry Pi. A *potentiometer*, or *pot*, is essentially a variable voltage divider, and can come in the form of a dial or slider.

The value of the potentiometer you use doesn't have to be 2K. If you only have a 10K or 1M potentiometer, it will work just fine.

To connect a potentiometer to the ADS1x15:

1. Insert the potentiometer into the breadboard.

2. The pot has three pins. Connect the middle pin to pin A0 on the ADS1x15.

3. One of the outside pins should connect to the positive rail of the breadboard. For now, it doesn't matter which.

4. Connect the other outside pin to the negative rail of the breadboard.

The connections should look as shown in Figure 8-4.

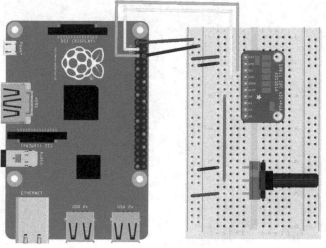

fritzing

Figure 8-4. *Using the ADS1x15 to connect a potentiometer to the Raspberry Pi*

Before you can read the potentiometer, you'll need to enable I^2C and install a few things (see Figure 8-5):

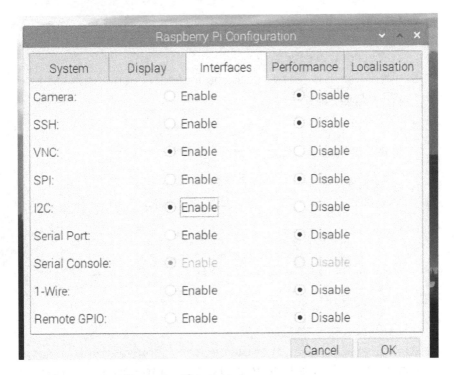

Figure 8-5. *Enabling the I²C interface*

1. Open up the Raspberry Pi Configuration tool by clicking on Menu→Preferences→Raspberry Pi Configuration.
2. Click the Interfaces tab.
3. Next to I²C , click Enable.
4. Click OK and reboot your Raspberry Pi.
5. Open up a Terminal window and update your list of packages:

   ```
   $ sudo apt-get update
   ```

6. The `i2c-tools` tools should already be installed with the latest version of the Raspberry Pi OS, but if you want to make sure, enter the following line at the command prompt. You'll either get a message that you've already got the newest version, or you will be prompted to install it.

   ```
   $ sudo apt-get install i2c-tools
   ```

7. Restart your Raspberry Pi.

8. After you've restarted your Raspberry Pi, test that the Raspberry Pi can detect the ADS1x15 with the command:

```
$ sudo i2cdetect -y 1
```

9. If the board is recognized, you'll see the number in the grid that is displayed:

```
     0  1  2  3  4  5  6  7  8  9  a  b  c  d  e  f
00:          -- -- -- -- -- -- -- -- -- -- -- --
10: -- -- -- -- -- -- -- -- -- -- -- -- -- -- -- --
20: -- -- -- -- -- -- -- -- -- -- -- -- -- -- -- --
30: -- -- -- -- -- -- -- -- -- -- -- -- -- -- -- --
40: -- -- -- -- -- -- -- -- 48 -- -- -- -- -- -- --
50: -- -- -- -- -- -- -- -- -- -- -- -- -- -- -- --
60: -- -- -- -- -- -- -- -- -- -- -- -- -- -- -- --
70: -- -- -- -- -- -- -- --
```

10. Now that we know that the device is connected and is recognized by our Pi, it's time to start reading the potentiometer. To do so, download the Raspberry Pi Python library from Adafruit's code repository into your home folder. Type the following command at the shell prompt, all on one line with nospaces in the URL:

```
wget https://github.com/adafruit/Adafruit_Python_ADS1x15/
    archive/master.zip
```

11. Unzip it:

```
$ unzip master.zip
```

12. Change to the library's ADS1x15 directory:

```
$ cd Adafruit_Python_ADS1x15-master/
```

13. Install the library:

```
$ sudo python3 setup.py install
```

14. Run one of the example files:

```
$ python3 examples/simpletest.py
```

15. Turn the potentiometer all the way in one direction and back in the other. Notice how the value in the pin 0 column changes:

```
Reading ADS1x15 values, press Ctrl-C to quit...

  |   0   |   1   |   2   |   3   |
  ------------------------------------
  |   0   | 4320  | 4496  | 4528  |
  |   0   | 4224  | 5104  | 4432  |

  |   0   | 4352  | 4688  | 5104  |
  | 6128  | 4640  | 4384  | 5552  |
  | 10928 | 4592  | 4592  | 5056  |
  | 18736 | 4384  | 4992  | 4384  |
  | 23312 | 4496  | 4912  | 4784  |
  | 26512 | 4912  | 4800  | 5968  |
  | 25968 | 4816  | 4800  | 5008  |
  | 16528 | 4416  | 4928  | 4928  |
  | 9280  | 4656  | 4416  | 5312  |
  | 2688  | 4240  | 5008  | 4464  |
  |   0   | 4384  | 4352  | 5360  |
  |   0   | 4336  | 4624  | 4272  |
  |   0   | 4256  | 4432  | 4592  |
```

16. Press Ctrl-C to terminate the script.

As you can see, turning the dial on the potentiometer changes the voltage coming into channel 0 of the ADS1x15. The code in the example does a little bit of math to determine the voltage value from the data coming from the ADC. Of course, your math will vary depending on what kind of sensor you use.

You can look inside that example script to see how it works, or try out the even simpler example that follows to learn how to take readings from the ADC. Create a new file with the code from Example 8-1 and execute it.

Example 8-1. Writing the code to read the ADC

```
from Adafruit_ADS1x15
from time import sleep   ❶

adc = Adafruit_ADS1x15.ADS1115()   ❷

while True:
    result = adc.read_adc(0)   ❸
    print result
    sleep(.5)
```

❶ Import Adafruit's ADS1x15 library.
❷ Create a new ADS1x15 object called adc.
❸ Get a reading from channel 0 on the ADS1x15 and store it in result.

When you run this code, it will output raw numbers for each reading twice a second. Turning the potentiometer will make the values go up or down.

Once you get it all set up, the Adafruit ADS1x15 library does all the hard work for you and makes it easy to use analog sensors in your projects. For instance, if you want to make your own Pong-like game, you could read two potentiometers and then use Pygame to draw on the game on the screen. For more information about using Pygame, see pygame.org.

Variable Resistors

Not all analog sensors work like the potentiometer, which provides a variable voltage based on some factor (such as the amount the dial on the pot is turned).

Some sensors are simply variable resistors that change the flow of electricity through the circuit, based on some factor. For instance, a photocell like the one in Figure 8-6 is a resistor that changes values based on the amount of light hitting the cell. Add more light, and the resistance goes down. Take away light, and the resistance goes up. On the other hand, a force-sensitive resistor decreases its resistance as you put pressure on the pad.

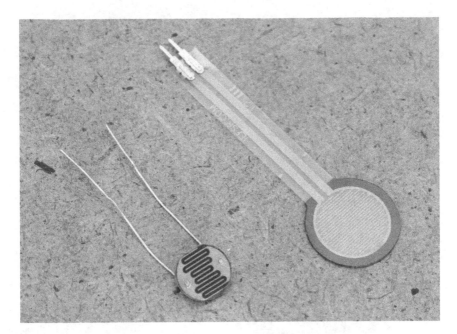

Figure 8-6. *The photocell and force-sensitive resistors act as variable resistors and can be used as analog inputs.*

To read sensors like these with analog input pins, you'll need to incorporate a *voltage divider* circuit.

Voltage Divider Circuit

When you're working with sensors that offer variable resistance, the purpose of a voltage divider is to convert the variable resistance into a variable voltage, which is what the analog input pins are measuring. First, take a look at a simple voltage divider.

In Figure 8-7, you'll see two resistors of the same value in series between the positive and ground. A single wire is connected to analog input 0, *between the two resistors*. Because both resistors are the same, 10KΩ, the voltage is divided neatly in half. Since the source voltage is 3.3V, there's about half of that, 1.65V, going to analog input 0.

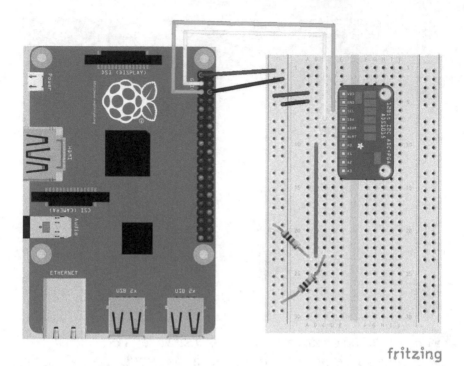

fritzing

Figure 8-7. *With two of the same value resistors between voltage and ground, the voltage between the two would be half of the total voltage.*

Without getting bogged down in the math involved, if you removed one 10K resistor and replaced it with a resistor of a higher value, the voltage going into the analog pin would *decrease.* If you removed that 10K resistor and replaced it with one of a lower value, the voltage going into the analog pin would *increase.* We can use this principle with sensors that are variable resistors to read them with analog input pins. You'll simply replace one of the resistors with your sensor.

To try the circuit out, you'll have to wire up a type of variable resistor called a *force-sensitive resistor*, or FSR.

Force-Sensitive Resistor

A force-sensitive resistor is a variable resistor that changes based on the amount of pressure placed on its pad. When there's no pressure on the pad, the circuit is open. When you start placing pressure on it, the resistance drops.

The exact figures will depend on your particular FSR, but typically you'll see 100KΩ of resistance with light pressure and 1Ω of resistance with maximum pressure. If you have a *multimeter*, you can measure the changes in resistance to see for yourself, or you can look at the component's *datasheet*, which will tell you what to expect from the sensor.

If you're going to replace the resistor connected to 3.3V in Figure 8-7 with a variable resistor like an FSR, you'll want the value of the other resistor to be somewhere in between the minimum and maximum resistance so that you can get the most range out of the sensor. For a typical FSR, try a 10KΩ resistor. To give the force-sensitive resistor a test drive:

1. Wire up an FSR to the ADC as shown in Figure 8-8.
2. Execute the code in Example 8-1 again.
3. Watch the readings on the screen as you squeeze the FSR.

If everything is working correctly, you should see the values rise as you increasingly squeeze harder on the FSR. As you press harder, you're reducing the resistance between the two leads on the FSR. This has the effect of sending higher voltage to the analog input.

You'll encounter many analog sensors that use this very same principle, and a simple voltage divider circuit, along with an analog-to-digital converter, will allow your Raspberry Pi to capture that sensor data.

fritzing

Figure 8-8. *Wiring up a force-sensitive resistor to the ADC requires a voltage divider circuit.*

Going Further

Controlling a DC Motor with PWM (bit.ly/1xpmBIF)
Adafruit has an excellent guide that teaches you how to use PWM to control the speed of a DC motor.

Reading Resistive Sensors with RC Timing (bit.ly/1rk6yJl)
Adafruit also shows you how to use a simple circuit to read resistive analog sensors without an analog-to-digital converter.

9/Working with Cameras

One of the advantages of using a platform like the Raspberry Pi for DIY technology projects is that it supports a wide range of peripheral devices. Not only can you hook up a keyboard and mouse, you can also connect accessories like printers, W-iFi adapters (at least on the first two generations of Raspberry Pi—the Model 3 and 4 have their own built-in), thumb drives, additional memory cards, cameras, and hard drives. In this chapter, we're going to show you a few ways to use a camera in your Raspberry Pi projects.

While not quite as common as a keyboard and mouse, a webcam is almost a standard peripheral for computers these days. Most laptops sold have a tiny camera built into the bezel of the display. And if they don't have a built-in camera, a USB webcam from a well-known brand can be purchased for as little as $25. You can even find webcams for much less if you take a chance on an unknown brand.

The folks at the Raspberry Pi Foundation have created their own camera peripheral that is designed to work with Raspberry Pi (Figure 9-1). Unlike a USB webcam, you're unlikely to find the official Raspberry Pi camera module in an office supply store, but you should be able to buy it wherever Raspberry Pis are sold, for around $25.

As of this writing, the newest version has a Sony IMX219 8-megapixel sensor (compared to the 5-megapixel OmniVision OV5647 sensor of the original camera). In addition, there is a new high-quality camera available for the Pi, with a 12.3 mega-pixel Sony IMX477 sensor, 7.9mm diagonal image size, and back-illuminated sensor architecture, with adjustable back focus and support for C- and CS-mount lenses.

And just in case those aren't quite cool enough options for you, you can also pick up an infrared camera—the Pi NoIR camera, with the same specs as the new version 2. This camera just doesn't have an infrared filter attached, so you're able to capture images using infrared lighting. Luckily, all of these different cameras use the same connector and the same code to access, so whichever version you're using, the code you see here will work with your camera.

Figure 9-1. *Raspberry Pi's camera module*

Unlike a USB webcam, the camera board connects to Raspberry Pi's Camera Serial Interface (CSI) connector (Figure 9-2). The reason is this: since the Broadcom chip at the core of the Raspberry Pi is meant for mobile phones and tablets, the CSI connection is how a mobile device manufacturer would connect a camera to the chip. Throughout this chapter, we'll use the official camera board as our chief example, but many of the projects and tutorials can also be done with a USB webcam (Figure 9-3).

Figure 9-2. *Raspberry Pi's camera serial interface*

Figure 9-3. *A typical USB webcam*

Connecting and Testing the Camera Module

Connecting the official camera module isn't as straightforward as connecting a USB device, but once you get it working, it should be a piece of cake.

 Make sure the Raspberry Pi is powered down before you do this.

Here are the steps you'll need to take:

1. Pull up on the edges of the CSI connector, which is right next to the Ethernet port. A piece of the connector will slide up and lean back toward the Ethernet port. (See Figure 9-4.)

Figure 9-4. *Opening the camera serial interface connector locking mechanism*

2. Insert the camera module's ribbon cable into the CSI connector so that its metal contacts are facing away from the USB ports.

3. Hold the ribbon cable into the CSI connector and press the moving part of the CSI connector down to lock and hold the ribbon cable in place. You'll still see part of the metal contacts on the ribbon cable.(See Figure 9-5.)

Figure 9-5. *After placing the ribbon cable into the camera serial interface connector and locking it down, you may still see metal contacts on the ribbon cable.*

4. Power up the Raspberry Pi and open the Configuration tool from the desktop Menu→Preferences. (See Figure 9-6.)

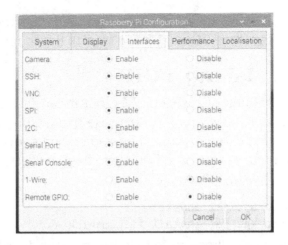

Figure 9-6. *Enabling the camera interface in the Raspberry Pi configuration tool*

5. Click the "Interfaces" tab.

6. Click the option to enable the camera.

7. Click OK and reboot your Raspberry Pi.

8. After you reboot, open a Terminal session and test the camera with:

```
$ raspistill -o test.jpg
```

If everything works, you'll see a preview image from the camera appear on the screen for a few seconds. After it disappears, a JPEG file of the captured image will be saved in your current directory. (An interesting side note here: if you're connecting to your Pi using the VNC connection and you run the command shown above in your terminal, the preview image will *not* appear in your VNC window, though it *will* appear on the Pi's main screen, should you have it connected to a monitor. This is because the `raspistill` command is interfacing with a different display than the VNC connection.)

`raspistill` is a powerful program with a lot of options. To see what's possible with it, view all the command-line options by running the program and piping the output through `less`:

```
$ raspi still 2>&1 | less
```

Use the up and down arrow keys to scroll through the options, and press q when you want to get back to the command line.

Project: Making a GIF

One of the features of `raspistill` is that it can take a series of photos at a specific interval. We can use this feature, along with the command-line image converting and editing software *Image-Magick*, to create fun animated GIFs with the Raspberry Pi:

1. First, install ImageMagick:

```
$ sudo apt-get update
$ sudo apt-get install imagemagick
```

2. Create a new directory to hold the images you capture and switch to that directory:

```
$ mkdir gif-test
$ cd gif-test
```

3. With your camera ready, execute `raspistill` to run for nine seconds, taking a 640×480 resolution image every three seconds, naming each file with an incrementing filename:

```
$ raspistill -t 9000 -tl 3000 -o image%02d.jpg -w
  640 -h 480
```

4. Next, input those files into ImageMagick outputting as thefile *test.gif*:

```
$ convert -delay 15 *.jpgtest .gif
```

5. Now open the *test.gif* by double-clicking the file within the desktop environment, and you'll see the GIF you made!

Capturing Video

There's also a command-line utility called `raspivid` to capture video from the official Raspberry Pi camera module. Try capturing a five-second video and saving it to a file:

```
$ raspivid -t 5000 -o video.h264
```

You can play that file back with:

```
$ omxplayer video.h264
```

which will open the video in the Pi's preinstalled VLC video player. And just like `raspistill`, `raspivid` is a powerful program with a lot of options. To see what's possible with it, view all the command-line options by running the program and piping the output through `less`:

```
$ raspivid 2>&1 less
```

Testing USB Webcams

With all the different models of webcams out there, there's no guarantee that a camera will work right out of the box. If you're purchasing a webcam for use with the Raspberry Pi, search online to make sure that others have had success with the model that you're purchasing. You can also check the webcam section of eLinux.org's page of peripherals (elinux.org/RPi_USB_Webcams) that have been verified to work with the Raspberry Pi.

Be aware that you may need to connect a powered USB hub to your Raspberry Pi if you want to connect your webcam in addition to your keyboard and mouse. The hub must be powered because the Raspberry Pi only lets a limited amount of electrical current through its USB ports, and it may not be able to provide enough power for your keyboard, mouse, *and* webcam—especially if you're using one of the older models of Pi. A powered USB hub plugs in to the wall and provides electrical current to the peripherals that connect to it so that they don't max out the power on your Raspberry Pi.

If you have a webcam that you're ready to test out with the Raspberry Pi, use `apt-get` in the Terminal to install a simple camera viewing application called luvcview:

```
$ sudo apt-get install luvcview
```

After `apt-get` finishes the installation, run the application by typing `luvcview` in a Terminal window while you're in the desktop environment. A window will open showing the view of the first video source it finds in the */dev* folder, likely */dev/video0*. Note the frame size that is printed in the Terminal window. If the video seems a little choppy, you can fix this by reducing the default size of the video. For example, if the default video size is 640×480, close luvcview and reopen it at half the video size by typing the following at the command line:

```
$ luvcview -s 320x240
```

If you still have your Pi camera connected when you connect the webcam, luvcview will show the Pi camera's view, since the Picamera is the source being pointed to by */dev/video0*. To test the webcam without disconnecting the Pi camera, type `luvcview -d /dev/video1` in the Terminal, and you should see the web-cam's viewpoint.

If you don't see video coming through, you'll want to troubleshoot here before moving on. One way to see what's wrong is by disconnecting the webcam, reconnecting it, and running the command `dmesg`, which will output diagnostic messages that might give you some clues as to what's wrong.

Installing and Testing OpenCV

To access the camera with Python code, we're going to be using OpenCV (Figure 9-7), which is a feature-packed open-source computer vision library. OpenCV makes it really easy to get images from the camera, display them on screen, or save them as files. But what makes OpenCV really stand out is its computer vision algorithms, which can do some pretty amazing things. Besides basic image transformations, it can also track, detect, and recognize objects in an image or video. Later on in this chapter, we'll try basic face detection with OpenCV ("Face Detection" on page 158). OpenCV is used by many working professionals in the fields of machine learning, artificial intelligence, and computer vision, and is considered a fully professional tool in these fields.

Figure 9-7. *Open CV logo*

To install OpenCV for Python, you'll need to start by installing the other libraries it depends on. For those, you can use apt-get:

```
$ sudo apt -get install libhdf5-dev libhdf5-serial-dev
libhdf5-103
$ sudo apt -get install libqtgui4 libqtwebkit4 libqt4-test
python3-pyqt5
$ sudo apt -get install libatlas-base-dev
$ sudo apt -get install libjasper-dev
```

It's a big install, and it may take a while before the process is complete.

Next, if you're starting with a fresh installation of the Raspberry Pi OS, you'll need to install `pip`, Python's package manager. To do that using wget (a command-line tool for interacting with and downloading files from the internet), enter the following into your Terminal:

```
$ wget https://bootstrap.pypa.io/get-pip.py
$ sudo python3 get-pip.py
```

Finally, you'll install the actual OpenCV library with the following command:

```
$ sudo pip install opencv-contrib-python==4.1.0.25
```

When it's done, check that the installation worked by going into the Python interactive interpreter and importing the library:

```
$ python3
Python 3.7.3 (default, Jan 22 2021, 20:04:44)
[GCC8.3.0] on linux
Type "help", "copyright", "credits" or "license" for more
    information.
>>> import cv2
>>> cv2.version
'4.1.0'
>>>
```

If you get no errors after importing the library and can list the library version, you know you've got OpenCV installed correctly. If you're using a USB webcam, you can jump ahead to "Displaying an Image" on page 151. If you're using the Raspberry Pi camera module, there may be one extra step, which we'll cover now.

Additional Step for the Raspberry Pi Camera Module

Because we want to use Python and OpenCV to access the Picamera module, we'll need to install one extra library—`picamera`. And because we're going to be using OpenCV, which makes use of NumPy arrays, we'll need to install the **array** sub-module. (The reason why I keep saying "may" is that depending on which version of the Raspberry Pi OS you're using, these modules may already be installed. If you're not sure, just do the following, and if the modules are installed, you'll simply get a message telling you so.) To install what you need, enter the following in theTerminal:

```
$ pip3 install "picamera[array]"
```

Now you're ready to interact with the camera using OpenCV!

Displaying an Image

For many of the examples in this chapter, you'll need to work in the desktop environment so that you can display images on the screen. You can work in IDLE, or save your code as *.py* files from the default text editor and execute them from the Terminal window. And while you *can* use a VNC connection, we don't recommend it, simply because you often won't be able to see the camera display (see the note above.)

We're going to start you off with some OpenCV basics using image files, and then you'll work your way up to reading images from the camera. Once you've got images coming in from the camera, it will be time to try some face detection:

1. Create a new directory within your home directory called *opencv-test*.

2. Open the web browser and search for an image that interests you. I used a photograph of raspberries from Wikipedia and re-named it *raspberries.jpeg*.

3. Right-click on the image and click "Save Image As."

4. Save the image within the *opencv-test* folder.

5. In the File Manager (on the Accessories menu), open the *opencv-test* folder and right-click in the folder. Choose Create New → Blank File.

6. Name the file *image-display.py*.

7. Double-click on the newly created *image-display.py* file to open it in the text editor.

8. Enter the code in Example 9-1.

9. Save the *image-display.py* file and run it from the Terminal window. If you've got everything right, you'll see a photo in a new

window as in Figure 9-8. You can close the window itself, or in the
Terminal, press Ctrl-C to end the script.

Figure 9-8. *The raspberry photo displayed in a window*

Example 9-1. Source code for image-display.py

```python
import cv2  ❶

img = cv2.imread("raspberries.jpeg")  ❷

cv2.imshow("Raspberries",img)  ❸

cv2.waitKey(0)  ❹
cv2.destroyAllWindows()  ❺
```

❶ Import the OpenCV library.
❷ Creates a new image object, `img`, and reads 'raspberries.jpeg'
    into it.
❸ Creates a window object named 'Raspberries' and displays the
    image object `img` within it.
❹ Wait for the user to press a key to end the program.
❺ Clean up after ourselves by destroying all windows when the
    program ends.

Image size

You may notice that the *image-display.py* script is an excellent example of a piece of code doing exactly what you tell it to—in this case, displaying an image in its exact form. If the image you choose happens to be 5440×2880 pixels, that's what OpenCV will display, regardless of whether or not it'll fit on your monitor. Keep reading to see how to resize your image to a more manageable display size.

Modifying an Image

Now that you can load an image into memory and display it on the screen, the next step is to modify the image before displaying it (doing this does not modify the image file itself; it simply modifies the copy of the image that's held in memory):

1. Save the *image-display.py* file as *superimpose.py*.

2. Make the enhancements to the code that are shown in Example 9-2.

3. Save the file and run it from the command line.

4. You should see the same image, but now superimposed with the shape and the text. To close the window, press 'Q' on your keyboard.

Example 9-2. Source code for superimpose.py

```
import cv2  ❶

img = cv2.imread("raspberries.jpeg'")

font = cv2.FONT_HERSHEY_SIMPLEX  ❷
org = (50,50)
font_scale = 1
color = (0, 255, 0)
thickness = 2

cv2.imshow("Raspberries", image)
```

```
cv2.rectangle(img, (30, 20), (270, 70), (255, 255, 255),
-1) ❸
image = cv2.putText(img, "Raspberries!", org, font,
font_scale, color, thickness,cv2.LINE_AA) ❹

cv2.imshow("Raspberries", image) ❺

cv2.waitKey(0)cv2.
destroyAllWindows()
```

❶ Import OpenCV.
❷ Specifications for the text: font, coordinates of the top left corner, scale, text size, and thickness
❸ On the image, draw a white rectangle from the coordinates (30, 20) to (270, 70) and fill it in.
❹ On the image (on top of the white rectangle), write the text "Raspberries!" using the preset specs.
❺ Display the image.

Figure 9-9. *The modified raspberry photo*

Instead of displaying the image on the screen, if you wanted to simply save your modifications to a file, Example 9-3 shows how the code would look.

Example 9-3. Source code for superimpose-save.py

```
import cv2
img = cv2.imread ('raspberries.jpeg')

font = cv2.FONT_HERSHEY_SIMPLEX
org = (50, 50)
font_scale = 1
color = (255, 0, 0)
thickness = 2

cv2.rectangle (img, (30, 20), (270, 70), (255, 255,
255), -1)
image = cv2.putText(img, 'Raspberries!', org, font,
font_scale, color, thickness, cv2.LINE_AA)

cv2.imwrite ('raspberries_text.jpeg', image)
```

Save the modified image in memory to a new file called *rasp-berries_text.jpeg*.

Because this code doesn't even open up a window, you can use it from the command line without the desktop environment running. You could even modify the code to watermark batches of images with a single command.

And you're not limited to text and rectangles. Here are a few of the other drawing functions available to you with OpenCV (their full documentation is available from OpenCV (docs.opencv.org/):
- Circle
- Ellipse
- Line
- Polygon
- Bezier curve

Accessing the Camera

Luckily, getting a camera's video stream into OpenCV isn't much different than accessing image files and loading them into memory. To try it out, you can make your own basic camera viewer:

1. Create a new file named *basic-camera.py* and save the code shown in Example 9-4 in it.

2. With your camera plugged in, run the script. You should see a window pop up with a view from the camera, as in Figure 9-10.

3. To close the window, press 'Q' on the keyboard.

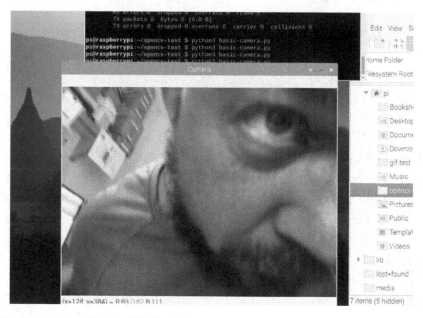

Figure 9-10. *Outputting camera input to the display*

Example 9-4. Source code for basic-camera.py

```
from pi camera.array import Pi RGBArray
from pi camera import PiCamera
import time
import cv2

camera = PiCamera()
```

```
camera.resolution = (640,480)  ❶
rawCapture = PiRGBArray(camera)

time.sleep (0.1)  ❷

camera.capture (rawCapture, format = 'bgr')  ❸
image = rawCapture.array

cv2.imshow ('Camera', image)  ❹
cv2.waitKey(0)
```

❶ Create a new camera object and set the height and width of the
    image to 640×480 for better performance and to fit the image
    on the screen better.
❷ Give the camera a tenth of a second to warm up.
❸ Get a frame from the camera, specifying "bgr" format rather
    than "rgb", because OpenCV uses BGR because it's just silly that
    way.
❹ Show the image in a window named, appropriately enough,
    "Camera".

You can even combine the code from the last two examples to make
a Python script that will take a picture from the camera and save
it as a *.jpeg* file:

```
from picamera.array import PiRGBArray
from picamera import PiCamera
import time
import cv2

camera = PiCamera()
camera.resolution = (640, 480)
rawCapture = PiRGBArray(camera)

time.sleep(0.1)

camera.capture(rawCapture,format='bgr')
image = rawCapture.array

cv2.imwrite('Camera.jpeg', image)
```

Face Detection

One of the most powerful functions that comes with OpenCV is built-in face detection. You can use deep learning or a few other algorithms, but the fastest and easiest way to detect faces is with *Haar cascades*. Haar cascades are basically pre-trained machine learning models, included in OpenCV, that allow you to detect faces with only a few lines of code. The process is a bit more error-prone than other methods, but it's faster and can run on resource-constrained machines such as a Pi. There are a few cascades included with OpenCV, such as face, nose, eye, mouth, and full body. Alternatively, you can download or generate your own cascade file if need be. `cv2.CascadeClassifier()` analyzes an image for matches, and if it finds at least one, the function returns the location of those matches within the image. This means that you can detect objects like cars, animals, or people within an image file or from the camera. To try out `CascadeClassifier()`, you can do some basic face detection:

1. First, you'll need to download the pre-trained classifier from OpenCV's Github page: github.com/opencv/opencv/blob/master/data/haarcascades/haarcascade_frontalface_default.xml. Download it, saving it under the same name, and store it in your *opencv-test* directory.

2. Find an image online that has at least one face in it; I'm using a publicly available image from morguefile.com.

3. Create a new file in the *opencv-test* directory called *face-detector.py*.

4. Enter the code shown in Example 9-5, changing the image name to match your downloaded image.

5. Run the script.

Example 9-5. Source code for face-detector.py

```
import cv2

face_cascade = cv2.CascadeClassifier('haarcascade_frontal
face_default.xml')  ❶
```

```
img = cv2.imread('faces.jpeg')

gray = cv2.cvtColor(img, cv2.COLOR_BGR2GRAY)  ❷

faces = face_cascade.detectMultiScale(gray, 1.1, 4)  ❸

for (x, y, w, h) in faces:
cv2.rectangle(img, (x, y),(x+w, y+h),(255, 0, 0),2)  ❹

cv2.imshow('Faces', img)
cv2.waitKey(0)
```

❶ Load the Haar cascades classifier.
❷ Convert the image to grayscale, as Haar cascades do not work on color images
❸ Detect faces according to the `detectMultiScale` function, which takes the image, the relative size of the included faces, and a parameter specifying how sensitive the algorithm should be to "face-like" structures.
❹ Draw a rectangle around each detected face in the image.

If your picture has faces in it but they're not being detected, try tweaking the last parameter of `detectMultiScale()` until you have some good results. You can also experiment with different images. As you can see, in the image I'm using for this test, there's a misidentified face in the middle of the photo. Playing around with that last parameter may help to avoid that misidentification. Also, this particular Haar cascade is meant to find faces that are in their normal orientation. If the faces in your photo are tilted or not fully facing the camera, this will affect the algorithm's ability to find them.

Figure 9-11. *Finding faces (mostly) in an image*

Project: Raspberry Pi Photobooth

You can combine different libraries to make Python a powerful tool to do some fairly complex projects. With the GPIO library you learned about in Chapter 7 and OpenCV, you can make your own Raspberry Pi-based photo booth that's sure to be a big hit at your next party (see Figure 9-12). And with the `CascadeClassifier()` function in OpenCV, you can enhance your photobooth with a special extra feature: the ability to automatically superimpose fun virtual props like hats, monocles, beards, and mustaches on the people in the photobooth. The code in this project is based on the code in the original editions of this book, which in turn is based on the Mustacheinator project in *Practical Computer Vision with SimpleCV* by Kurt Demaagd, Anthony Oliver, Nathan Oostendorp, and Katherine Scott (O'Reilly).

Figure 9-12. *Output of the Raspberry Pi Photobooth*

Here's what you'll need to turn your Raspberry Pi into a photobooth:

- A USB webcam or Raspberry Pi Camera Module
- A monitor
- A pushbutton, any kind you like
- Hook up wire, cut to size

Before you get started, make sure that both the gpiozero and OpenCV Python libraries are installed and working properly on your Raspberry Pi.

1. As you did in Chapter 7, connect pin 24 to the pushbutton. One side of the button should be connected to ground, the other to pin 24. (Remember, we're talking about BCM numbers here, not physical pins. You're actually connecting to physical pin #18 on the Pi.)

2. Find or create a small image of a black mustache on a white background and save it as *moustache.png* in a new folder called *photobooth* on your Raspberry Pi. You can also download a premade mustache file from the *images* subdirectory of the Github repository for this book: github.com/wdonat/gsw_raspi_4e.

3. From the same repository, grab the `haarcascade_mcs_mouth.xml` file and place it into the *photobooth* directory. (If you download the file from Github's web interface, the best way to do it is to get the raw version of the file, copy all of the text, and then paste it into a blank text document on your computer. Then save as `haarcascade_mcs_mouth.xml`.)

4. In the *photobooth* directory, create a new file called *photo-booth.py*, type in the code listed in Example 9-6, and save the file.

Example 9-6. Source code for photobooth.py

```
import cv2
from picamera.array import PiRGB Array
from picamera import PiCamera
from gpiozero import Button
import time

button = Button(24)
camera = PiCamera()
camera.resolution = (800, 608)  ❶
rawCapture = PiRGBArray (camera)

mouth_cascade =

cv2.CascadeClassifier(''haarcascade_mcs_mouth.xml')  ❷
```

```
moustache = cv2.imread('moustache.png')
rows,cols,_ = moustache.shape

moustache2gray = cv2.cvtColor(moustache,cv2.COLOR_
BGR2GRAY)
ret, mask = cv2.threshold(moustache2gray,10,255,cv2.
THRESH_BINARY)
mask_inv = cv2.bitwise_not(mask)  ❸

whileTrue:
    button.wait_for_press()
    camera.capture(rawCapture,format= 'bgr')
    cap = rawCapture.array
    cv2.imwrite('face.jpg', cap)  ❹
    image = cv2.imread('face.jpg')
    gray = cv2.cvtColor(image, cv2.COLOR_BGR2GRAY)
    mouths = mouth_cascade.detectMultiScale(gray, 1.5, 15)
❺

    for(x, y, w, h)inmouths:
        roi = image[y-rows+10:y+10, x-5:x+cols-5]  ❻

        face_bg = cv2.bitwise_and(roi, roi, mask = mask)  ❼
        moustache_fg = cv2.bitwise_and(moustache, mous-
tache, mask = mask_inv)  ❽
        dst = cv2.add(face_bg, moustache_fg)  ❾
        image[y-rows+10:y+10, x-5:x+cols-5] = dst  ❿

cv2.imshow('Photobooth', image)
cv2.waitKey(0)
cv2.destroyAllWindows()
time.sleep(0.1)
```

❶ Set the camera resolution. It's not *exactly* 800×600—hence
the 608 value. If you force the camera to use 600px, the image
often gets a strange blue cast, and we're not going for the An-
dorian or the *Avatar* look here.

❷ This is the Haar cascade that will search the image for a mouth.

❸ This line and the previous one remove everything but the black pixels in the mustache image and create a mask and an inverted mask.

❹ After capturing the image, you must write it to a file and then reopen it for editing. The raw image fresh from the capture is non-editable, even by sudo.

❺ This finds the mouth(s) in the grayscale image, using the Haar cascade.

❻ This determines a *region of interest* (ROI) above the mouth where we are going to place the mustache.

❼ This masks the face with the mustache image.

❽ This masks the non-mustache portion of the mustache image with the portion of the face image that lies within the ROI.

❾ This line literally adds the two images together.

❿ Finally, replace the ROI of the face image with the mustache + face image we just created.

Now you're ready to give it a try. Make sure your camera is connected. Next, go to the terminal, change to the *photobooth* directory, and then run the script:

```
$ python3 photobooth.py
```

Point the camera toward your face and press the button. If all of the parameters are correct and you're pointing the camera in the right direction, you'll see an image on the screen similar to the one above.

TROUBLESHOOTING:

This is where some serious experimentation on your part maybe necessary. The mouth classifier is not nearly as well-trained as the face classifier we used earlier, so you're bound to get either some false positives or the program won't recognize any mouths in the image at all, even if you're grinning from ear to ear at the camera. If either of these things happen, play around with the values in ❺ above. Also, make sure your face is well-lit, the camera is pointed directly at you, and your face is centered in the frame.

Next, there's the placement of the mustache. I've sized the one we're using here to fit pretty well on a well-centered face inside an 800×600 image, but you will most likely need to play with its positioning. To do that, experiment with the constants in ❻ and ❿ above. When you do, make sure that any changes you make to the first line are mirrored in the second line, or your script will probably fail with an error. (If you want to move the mustache to the left or right, change the y associated constant; up or down, change the x constant. It's a bit counter-intuitive.)

You can also play with different size images and see if they give you any better results.

Going Further

PyImage Search

Any work with OpenCV and the Raspberry Pi has to take into account Dr. Adrian Rosebrock's excellent PyImage Search site. This site is exceptionally well done and has a wealth of information about almost anything you care to learn about using OpenCV, whether on Linux or the Pi or even other operating systems. In addition, he's been branching out into machine learning, neural networks, and AI. Definitely worth your time to peruse and bookmark.

OpenCV Documentation

As always, checking out the online documentation for OpenCV is never a bad idea. In my opinion, the docs for OpenCV are not very well indexed or user-friendly, but there is a *lot* of information there and it may be worthwhile to browse if you have questions or even if you're just curious as to everything OpenCV can do.

10/Python and the Internet

Python has a very active community of de-velopers who often share their work in the form of open-source libraries that simpli-fy complex tasks.some of these libraries make it relatively easy for us to connect our projects to the internet to do things like get-ting data about the weather, send an email or text message, follow trends on Twitter, or act as a webserver.

In this chapter, we're going to take a look at a few ways to create internet-connected projects with the Raspberry Pi. We'll start by showing you how to fetch data from the internet and then move into how you can create your own Raspberry Pi-based webserver.

Download Data from a Web Server

When you type an address into your web browser and hit Enter, your browser is the *client*. It establishes a connection with the *server*, which responds with a web page. Of course, a client doesn't have to be a web browser; it can also be a mail application, a weather wid-get on your phone or computer, or a game that uploads your high score to a global leaderboard. In the first part of this chapter, we're going to focus on projects that use the Raspberry Pi to act as a cli-ent. The code you'll be using will connect to internet servers to get information. Before you can do that, you'll need to install a popular Python library called Requests that is used for connecting to web servers via *hyper-text transfer protocol*, or HTTP.

To use Requests in Python, you first need to import it. Within Terminal:

```
$ python3
Python3.7.3(default, Jan222021, 20:04:44)
[GCC8.3.0]onlinux
Type "help", "copyright", "credits" or "license" for
more information.
>>> import requests
>>>
```

If you don't get any kind of error message, you'll know Requests has been imported in this Python session.

Now you can try it out:

```
>>> r = requests.get('http://www.google.com/')
>>>
```

You may be a bit disappointed at first because it seems like nothing happened. But actually, all the data from the request has been stored in the object r. Here's how you can display the status code:

```
>>> r.status_code
200
```

The HTTP status code 200 means that the request succeeded. There are a few other HTTP status codes inTable 10-1.

Table 10-1. *Common HTTP status codes*

Code	Meaning
200	OK
301	Moved permanently
307	Moved temporarily
401	Unauthorized
404	Not found
500	Server error

If you want to see the contents of the *response* (what the server sends back to you), try the following:

```
>>> r.text
```

If everything worked correctly, what will follow is a large block of text; you may notice some human-readable bits in there, but most of it will be hard to understand. This is the raw HTML of Google's landing page, which is meant to be interpreted and rendered on-screen by a web browser.

However, not all HTTP requests are meant to be rendered by a web browser. Sometimes only data is transmitted, with no information about how it should be displayed. Many sites make these data protocols available to the public so that we can use them to fetch data from and send data to their servers without using a web browser. Such a data protocol specification is commonly called an *application programming interface*, or API. APIs let different pieces of software talk to each other and are popular for sending data from one site to another over the Internet.

For example, let's say you want to make a project that will sit by your door and remind you to take your umbrella with you when rain is expected that day. Instead of setting up your own weather station and figuring out how to forecast the precipitation, you can get the day's forecast from one of many weather APIs out there.

Fetching the Weather Forecast

To determine whether or not it will rain today, we'll show you how to use the API from Weather Underground (www.wunderground. com).

To use the API, take the following steps:

1. In a web browser, go to Weather Underground's API homepage (www.wunderground.com/weather/api) and enter your information to sign up.

2. After signing up, you'll need to create and register a weather station to be issued an API key. Don't sweat it—you can register a Raspberry Pi as a weather station.

3. Once you've signed up, go back to weather/api (www.wunderground.com/weather/api). If this is your first login, you'll see a notice that you don't have any API keys assigned yet because you must own a weather station. Click the "Learn More" button

to be taken to the Personal Weather Station Network page, and then click the "Register" tab.

4. Click "Add New Device" (Figure 10-1).

5. On the Add New Device page, choose Raspberry Pi from the drop down menu under Personal Weather Station and click "Next".

6. Now enter your home address (or wherever you want to use your Pi weather station) and click "Next". On the next screen, give your device a name and enter a height above the ground at which your device will be stored. Accept the Privacy Agreement and click "Next."

7. At this point your registration is complete (figure 10-2).Copy the credentials down, as you'll need them later, and then click the "My Devices" button. Finally, on *that* page, click the API Keys tab. Check the Terms of Service checkbox and click "Generate New Key". Your new API key will be generated and populated into the text box. You should copy it somewhere and save it.

8. From here, if you click the "View API Documentation" button, you'll be taken to a Google docs page detailing the APIs that are available to you. To learn more about how to use any of them, click on its associated link at the right. For this project, click on the "Forecast" link at the bottom.

9. This will take you to another Google doc page with five pages of information about the fields available, what they each mean, different API calls you can make, the parameters required for each, and finally an example of a JSON response to one of the API calls. Take a look at the first line, which shows the URL needed to get the weather forecast at any location. Note that embedded in the URL is your API key, as well as the latitude and longitude of a particular weather station (I'm not sure which one). The API key is not filled in in the example, but the latitude and longitude are. If you want to see the forecast for any place on earth, you can use your own API key, and just change the location in the URL.

10. To try out the forecast API, copy the URL to your browser address bar, substituting your API key where required. You should see the forecast data in a format called JSON, or Java Script Object Notation (see Example 10-1). Congratulations! You've just interacted with the Weather Underground's API!

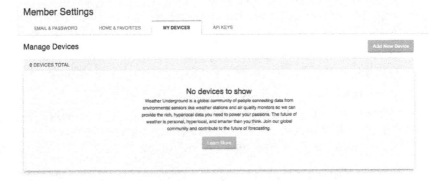

Figure 10-1. *Adding a new device*

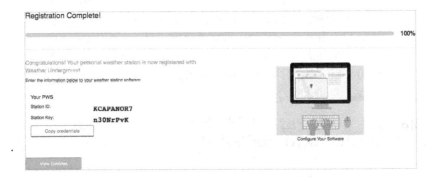

Figure 10-2. *Registration complete*

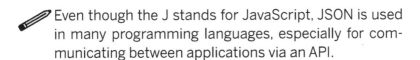 Even though the J stands for JavaScript, JSON is used in many programming languages, especially for communicating between applications via an API.

When I'm first interacting with an API and I want to make sense of its responses, sometimes I'll copy the text to a text editor and break it down, line by line, to make it more human-readable. This way I can see the natural hierarchy of the JSON and figure out exactly what I have to look for when I'm decoding the response in my code. It starts with putting the key: value pairs on the individual lines (example 10-2). Here you can see that there are keys like "sunriseTime Local", and "precipChance", and "narrative", which is the verbal description of the forecast for the coming days of the week.

Keep in mind that not all APIs are created equal and you'll have to review their documentation to determine if it's the right one for your project. Also, most APIs limit the number of requests you can make, and some even charge you to use their services. Many times, the API providers have a free tier for a small number of daily requests, which is perfect for experimentation and personal use.

Example 10-1. Partial JSON response from WeatherUnderground'sAPI

```
{
    "calendarDayTemperatureMax": [74,83,75,73,62,63],
    "calendarDayTemperatureMin": [49,55,63,59,52,50],
    "dayOfWeek": ["Saturday","Sunday","Monday","Tuesday",
"Wednesday","Thursday"]
    ...
    "daypart": [
        {
            "cloudCover": [null,37,38,78,86,65,79,89,92,82,83,61],
            "dayOrNight":
[null,"N","D","N","D","N","D","N","D","N","D","N"],
            "precipChance": [null,1,1,49,51,24,24,56,55,39,33,2
4],
            ....
        }
    ]
}
```

As it happens, JSON is identical in form to Python's dictionary data type (a set of key/value entries), which makes it very easy to write Python code that reads and writes JSON to interact with various APIs. To do this, you can import Python's json library.

If you would like to really get a sense of what the API offers, you can make some commands in your terminal:

```
$ python3

>>> import requests
>>> key ='<YOUR KEY HERE>"
>>> api_url = 'https://api.weather.com/v3/wx/forecast/dai-
ly/
5day?geocode=33.74,-84.39&format=json&units=e&language=en-
US&apiKey=' + key
>>> r = requests.get(api_url)
>>> forecast = r.json()
>>> for key in forecast:
...     print(key)
```

(Note that when you're interacting with Python in the terminal this way, command-by-command, after entering a command that requires an indented block of code such as for, use the TAB key for each indented line that follows. You'll see the three-dot ellipse as evidence of the indentation. Hit return once again to end the indented block.)

These commands will return

```
calendarDayTemperatureMax
calendarDayTemperatureMin
dayOfWeek
expirationTimeUtc
moonPhase
moonPhaseCode
moonPhaseDay
moonriseTimeLocal
moonriseTimeUtc
moonsetTimeLocal
moonsetTimeUtc
narrative
qpf
qpfSnow
```

```
sunriseTimeLocal
sunriseTimeUtc
sunsetTimeLocal
sunsetTimeUtc
temperatureMax
temperatureMin
validTimeLocal
validTimeUtcdaypart
```

so you can see exactly what keys are available to look at. Note that 'daypart' is a one-item list that contains a dictionary, so you can dig a little deeper with

```
>>> for key in forecast ['daypart'][0]:
...    print(key)
```

This will return

```
cloudCover
dayOrNight
daypartName
iconCode
iconCodeExtend
narrativeprecip
ChanceprecipType
qpf
qpfSnow
qualifierCode
qualifierPhrase
relativeHumidity
snowRange
temperature
temperatureHeatIndex
temperatureWindChill
thunderCategory
thunderIndex
uvDescription
uvIndex
windDirection
windDirectionCardinal
windPhrase
windSpeed
wxPhraseLong
wxPhraseShort
```

As you can see, there's a lot of information packed into this one simple API call, and you can spend a lot of time with it to get the information you want. But if you remember, we wanted a script that would tell us if we need an umbrella. Because there's a key inside 'daypart' called precipChance, we can just query that key's value to see the percent chance that it will rain that day. For a rain forecast indicator, let's say that any probability of precipitation value over 30% is a day that we want to have an umbrella handy.

1. Connect an LED to pin 25, as you did in Figure 6-5.

2. Create a new file called *umbrella-indicator.py* and use the code in Example 10-2. Don't forget to put in your own API key and the location in the Weather Underground API URL.

3. Run the script as root with the command sudo python3 umbrella-indicator.py.

Example 10-2. Source code for umbrella-indicator.py

```
import requests
import time
from gpiozero import LED

key = '<YOUR API KEY HERE>'  ❶
latitude = '<YOUR LATITUDE>' Use quotes to make it a string
longitude = '<YOUR LONGITUDE>' # Again, use quotes
api_url = 'https://api.weather.com/v3/wx/forecast/daily/5day?
geocode=' + latitude +', ' + longitude +
'&format=json&units=e&language=en-US&apiKey=' + key

while True:
    r = requests.get(api_url)
    forecast = r.json()
    pop_value = forecast['daypart'][0]['precipChance']  ❷
    if pop_value is None:
        pop_value = 0  ❸
    if pop_value >= 30:  ❹
        led.on()
    else:
        led.off()  ❺
    time.sleep(180) # 3minutes  ❻
```

❶ As before, change this to your API key.

❷ Get today's probability of precipitation and store it in `popValue`.

❸ Convert `popValue` from a string into an integer so that we can evaluate it as a number.

❹ If the value is greater than 30, then turn the LED on.

❺ Otherwise, turn the LED off.

❻ Wait three minutes before checking again so that the script stays within the API limit of 500 requests per day.

As you may have already discovered, there's a small sticking point in the API that we need to take into consideration in our script: if there is no chance of precipitation on a particular day, the value saved into the `precipChance` list is "None" This means that if we check for the value of `precipChance` on that day, instead of getting "0" back as a result, we'll get None, which is literally "no value". We can't compare "no value" to 30, so the script will fail if we try to run it on a day with no precipitation in the forecast. To account for that failure, if `popValue` is None, we set it to zero instead and go on about our business.

Press Ctrl-C to quit the program when you're done.

The Weather Underground API is one of a plethora of different APIs that you can experiment with. Table 10-2 lists a few other sites and services that have APIs.

Table 10-2. *Popular application programming interfaces*

Site	API Reference URL
Facebook	developers.facebook.com
Flickr	www.flickr.com/services/api
Four-square	developer.foursquare.com
Reddit	www.reddit.com/dev/api
Twilio	www.twilio.com
Twitter	dev.twitter.com
YouTube	developers.google.com/youtube

Serving Pi (Be a Web Server)

Not only can you use the Raspberry Pi to get data from servers via the internet, but your Pi can also act as a server itself. There are many different web servers that you can install on the Raspberry Pi. Traditional web servers, like Apache or Lighttpd, serve the files from your board to clients. Most of the time, servers like these are sending HTML files and images to make web pages, but they can also serve sound, video, executable programs, and much more.

However, there's a new breed of tools that extend programming languages like Python, Ruby, and JavaScript to create web servers that dynamically generate the HTML when they receive HTTP requests from a web browser. This is a great way to trigger physical events, store data, or check the value of a sensor remotely via a web browser. You can even create your own JSON API for an electronics project!

Flask Basics

We're going to use a Python web framework called Flask (flask. palletsprojects.com) to turn the Raspberry Pi into a dynamic web server. While there's a lot you can do with Flask "out of the box," it also supports many different extensions for doing things, such as user authentication, generating forms, and using databases. You also have access to the wide variety of standard Python libraries that are available to you.

Why Flask and not Django?

Flask and Django are two different frameworks for developing web stuff in Python. (A framework is a collection of modules or packages that help a coder to write apps or services without having to worry about low-level details.) Django is a fine framework for developing web apps in Python. But for our purposes, Flask is better: it is smaller, with fewer restrictions for the coder, and fits the needs of this chapter beautifully.

Here's how to install Flask and its dependencies:

```
$ pip3 install --user flask
```

To test the installation, create a new file called *hello-flask.py* with the code from Example 10-3. Don't worry if it looks a bit overwhelming at first; you don't need to understand what every line of code means right now. The block of code that's most important is the one that contains the string "Hello World!"

Example 10-3. Source code for hello-flask.py

```
from flask import Flask
app = Flask(__name__)  ❶

@app.route("/")  ❷
def hello():
    return "Hello World!"  ❸

if __name__ == "__main__":  ❹
    app.run(host='0.0.0.0', port=80, debug=True)  ❺
```

❶ Create a Flask object called **app**.
❷ Run the code below when someone accesses the root URL of the server.
❸ Send the text "Hello World!" to the client.
❹ If this script was run directly from the command line.
❺ Have the server listen on port 80 and report any errors.

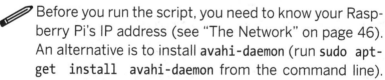 Before you run the script, you need to know your Raspberry Pi's IP address (see "The Network" on page 46). An alternative is to install **avahi-daemon** (run **sudo apt-get install avahi-daemon** from the command line). This lets you access the Pi on your local network through the address raspberrypi.local. If you're accessing the Raspberry Pi webserver from a Windows machine, you may need to also put Bonjour Services (bit.ly/1sjViwr) on it for this to work.

Now you're ready to run the server, which you'll have to do as root:

```
$ sudo python3 hello-flask.py
 * Serving Flask app "hello-flask" (lazy loading)
 * Environment:production
WARNING: Do not use the development server in a produc-
tion environment. Use a production WSGI server instead.
 * Debug mode: on
 * Running on http://0.0.0.0:80/ (Press CTRL+C to quit)
 * Restarting with stat
 * Debugger is active!
 * Debugger PIN:328-390-680
```

From another computer on the same network as the Raspberry Pi, type your Raspberry Pi's IP address into a web browser. If your browser displays "Hello World!", you know you've got it configured-correctly. You may also notice that a few lines appear in the terminal of the Raspberry Pi:

```
192.68.2.4 - - [09/May/202117:16:31] "GET/HTTP/1.1" 200 -
192.68.2.4 - - [09/May/202100:31:31] "GET/favicon.
icoHTTP/
   1.1" 404 -
```

The first line shows that the web browser requested the root URL and our server returned HTTP status code 200 for "OK." The second line is a request that many web browsers send automatically to get a small icon called a *favicon* to display next to the URL in the browser's address bar. Our server doesn't have a *favicon.ico* file, so it returned HTTP status code 404 to indicate that the URL was not found.

If you want to send the browser a site formatted in proper HTML, it doesn't make a lot of sense to put all the HTML into your Python script. Flask uses a template engine called Jinja2 (jinja. pocoo.org/docs/templates) so that you can use separate HTML files that contain placeholders where you want dynamic data to be inserted.

If you've still got *hello-flask.py* running, press Ctrl-C to kill it.

To make a template, create a new file called *hello-template.py* with the code from Example 10-4. In the same directory with *hello-template.py*, create a subdirectory called *templates*. In the *templates* subdirectory, create a file called *main.html* and insert the code from Example 10-5. Anything in double curly braces within the HTML template is interpreted as a variable that would be passed to it from the Python script via the render_template function.

Example 10-4. Source code for hello-template.py

```python
from flask import Flask, render_template
import datetime
app = Flask(__name__)

@app.route("/")
def hello():
    now = datetime.datetime.now()          ❶
    timeString = now.strftime("%Y-%m-%d%H:%M")   ❷
    templateData = {
        'title':'HELLO!',     ❸
        'time':timeString
        }
    return render_template('main.html', **templateData)  ❹

if __name__ == "__main__":
    app.run(host='0.0.0.0', port=80, debug=True)
```

❶ Get the current time and store it in now.

❷ Create a formatted string using the date and time from the now object.

❸ Create a *dictionary* of variables (a set of *keys*, such as title, that are associated with values, such as HELLO!) to pass into the template.

❹ Return the *main.html* template to the web browser using the variables in the templateData dictionary.

Example 10-5. Source code for templates/main.html

```html
<!DOCTYPE html>
<head>
    <title>{{ title }}</title>    ❶
</head>
<body>
    <h1>Hello, World!</h1>
    <h2>The date and time on the server is:{{ time }}</h2>   ❷
</body>
</html>
```

❶ Use the `title` variable in the HTML title of the site.
❷ Use the `time` variable on the page.

Now, when you run *hello-template.py* (as before, you need to use sudo to run it) and pull up your Raspberry Pi's address in your web browser, you should see a formatted HTML page with the title "HELLO!" and the Raspberry Pi's current date and time.

 It's unlikely that this page will be accessible from outside your local network via the internet, though that depends on how your network is set up. If you'd like to make the page available from outside your local network, you'll need to configure your router for port forwarding. When you do this, you'll be telling the router that incoming requests to its external IP address on port 80 (for example) should be sent directly to the Pi to be handled. Refer to your router's documentation for more information about how to do this.

Connecting the Web to the Real World

You can use Flask with other Python libraries to bring additional functionality to your site. For example, with the GPIO Zero Python module (see Chapter 7), you can create a website that interfaces with the physical world. To try it out, hook up three buttons or switches to pins 23, 24, and 25 in the same way as the Simple Soundboard project in Figure 7-2.

The following code expands the functionality of *hello-template.py*, so copy it to a new file called *hello-gpio.py*. Add the gpiozero module and a new *route* for reading the buttons, as we've done in Example 10-6. The new route will take a variable from the requested URL and use that to determine which pin to read.

Example 10-6. Modified source code for hello-gpio.py

```python
from flask import Flask, render_template
import datetime
from gpiozero import Button
app = Flask(__name__)

button1 = Button(24)
button2 = Button(25)
button3 = Button(26)

@app.route("/")
def hello():
    now = datetime.datetime.now()
    timeString = now.strftime("%Y-%m-%d%H:%M")
    templateData = {
        'title' : 'HELLO!',
        'time' : timeString
        }
    return render_template('main.html', **templateData)

@app.route("/readPin/<pin>")
def readPin(pin)  ❶
    try:  ❷
        if pin == '23':  ❸
            if button1.is_pressed:  ❹
                response = "Pin number 23 is high!"
            else:  ❺
                response = "Pin number 23 is low!"
        elif pin == '24':
            if button2.is_pressed:
                response = "Pin number 24 is high!"
            else:
                response = "Pin number 24 is low!"
        elif pin == '25':
            if button3.is_pressed:
                response = "Pin number 25 is high!"
            else:
                response = "Pin number 25 is low!"
    except:  ❻
        response = "There was an error reading pin" + pin + "."

    templateData = {
        'title' : 'StatusofPin' + pin,
```

```
'response' : response
}

return render_template('pin.html', **templateData)

if __name__ == "__main__":
app.run(host='0.0.0.0', port=80, debug=True)
```

❶ Add a dynamic route with pin number as a variable.

❷ If the code indented below raises an exception, run the code in the except Block.

❸ Take the pin number from the URL and check for the corresponding button (pin).

❹ If the pin is high, set the response text to say that it's high.

❺ Otherwise, set the response text to say that it's low.

❻ If there was an error reading the pin, set the response to indicate that.

You'll also need to create a new template called *pin.html* (it's not very different from *main.html*, so you may want to copy *main.html* to *pin.html* and make the appropriate changes, as in Example 10-7):

Example 10-7. Source code for templates/ pin.html

```
<!DOCTYPEhtml>
    <head>
        <title>{{ title }}</title>   ❶
    </head>

    <body>
        <h1>Pin Status</h1>   ❷
        <h2>{{ response }}</h2>
    </body>
</html>
```

❶ Insert the title provided from *hello-gpio.py* into the page's title.

❷ Place the response from *hello-gpio.py* on the page inside HTML heading tags.

With this script running, when you point your web browser to your Raspberry Pi's IP address, you should see the standard "Hello World!" page we created before. But add */readPin/24* to the end of the URL so that it looks something like 10.0.1.103/readPin/ 24. A page should display showing that the pin is being read as low. Now hold down the button connected to pin 24 and refresh the page; it should now show up as high!

Try the other buttons as well by changing the URL. The great part about this code is that we only had to write the function to read the pin once and create the HTML page once, but it's almost as though there are separate web pages for each of the pins!

Project: WebLamp

In Chapter 6, we showed you how to use Raspberry Pi as a simple AC outlet timer in "Project: Cron Lamp Timer" on page 100. Now that you know how to use Python and Flask, you can now control the state of a lamp over the Web. This basic project is simply a starting point for creating internet-connected devices with the Raspberry Pi.

And just as the previous Flask example showed how you can have the same code work on multiple pins, you'll setup this project so that if you want to control more devices in the future, they will be easy to add:

1. The hardware setup for this project is exactly the same as the "Project: Cron Lamp Timer" on page 100, so all the parts you need are listed there.

2. Connect the PowerSwitch Tail II relay to pin 25, just as you did in the Cron Lamp Timer project.

3. If you have another PowerSwitch Tail II relay, connect it to pin 24 to control a second AC device. Otherwise, just connect an LED to pin 24. We're simply using it to demonstrate how multiple devices can be controlled with the same code.

4. Create a new directory in your home directory called *Web-Lamp*.

5. In *WebLamp*, create a file called *weblamp.py* and put in the code from Example 10-8.

6. Create a new directory within *WebLamp* called *templates*.

7. Inside *templates*, create the file *main.html*. The source code of this file can be found in Example 10-9. In the terminal, navigate to the *WebLamp* directory and start the server. Be sure to use Ctrl-C to kill any other Flask server you have running first, because if you already have a server running, your code will not run. Instead, you'll get an error message that the address is already in use and the script will fail.

```
pi@raspberrypi ~/WebLamp $ sudo python3 weblamp.py
```

Open your mobile phone's web browser and enter your Raspberry Pi's IP address in the address bar, as shown in Figure 10-3.

Figure 10-3. *The device interface, as viewed through a mobile browser*

Example 10-8. Source code for weblamp.py

```python
from gpiozero import LED
from flask importFlask, render_template, request
app = Flask(__name__)

pins = {
    24 :{'name': 'coffeemaker', 'state': False},
    25 :{'name': 'lamp', 'state': False}
    } ❶

led24 = LED(24) ❷
led25 = LED(25)
led24.off()
led25.off()

@app.route("/") ❸
defmain():
    pins[24]['state'] = led24.is_lit
    pins[25]['state'] = led25.is_lit

    templateData = { ❹
       'pins' : pins
       }
    return render_template('main.html', **templateData) ❺

@app.route("/<changePin>/<action>") ❻
def action(changePin, action):
    changePin = int(changePin) ❼
    deviceName = pins[changePin]['name'] ❽
    if action == "on": ❾
       if changePin == 24: ❿
          led24.on()
          pins[24]['state'] = led24.is_lit
       if changePin == 25:
          led25.on()
          pins[25]['state'] = led25.is_lit
       message = "Turned" + deviceName + "on." ⓫

    if action == "off":
       if changePin == 24:
          led24.off()
          pins[24]['state'] = led24.is_lit
```

```
        if change Pin == 25:
            led25.off()
            pins[25]['state']=led25.is_lit
        message = "Turned" + deviceName + "off."

    if action == "toggle":
        if changePin == 24:
            led24.toggle()
            pins[24]['state'] = led24.is_lit
        if changePin == 25:
            led25.toggle()  ⓬
            pins[25]['state'] = led25.is_lit
        message = "Toggled" + deviceName + "."

    templateData = {  ⓭
        'message': message,
        'pins': pins
    }  ⓮
    returnrender_template('main.html', **templateData)
if __name__ == "__main__":
    app.run(host='0.0.0.0', port=80, debug=True)
```

❶ Create a dictionary called pins to store the pin number, name, and pin state.

❷ Set each pin as an output and make it low.

❸ For each pin, read the pin state and store it in the pins dictionary.

❹ Put the pins dictionary into the template data dictionary.

❺ Pass the template data into the template *main.html* and return it to the user.

❻ The function below is executed when someone requests a URL with the pin number and action in it.

❼ Convert the pin from the URL into an integer.

❽ Get the device name for the pin being changed.

❾ If the action part of the URL is "on," execute the code indented below.

❿ Set the pin high.

⓫ Save the status message to be passed into the template.

⓬ Read the pin and set it to whatever it isn't (i.e., toggle it).

- ⑬ For each pin, read the pin state and store it in the pins dictionary.
- ⑭ Along with the `pins` dictionary, put the message into the template data dictionary.

Example 10-9. Source code for templates/main. html

```
<!DOCTYPEhtml>
<head>
    <title>Current Status</title>
</head>

<body>
<h1>Device Listing and Status</h1>

    {% for pin in pin s%}  ❶
    <p>The {{ pins[pin].name }}  ❷
    {% if pins[pin].state == True %}  ❸
        is currently on (<a href="/{{pin}}/off">turn off</
        a>)
    {% else %}  ❹
        is currently off(<a href="/{{pin}}/on">turn on</a>)
    {% end if %}
    </p>
    {% end for %}

    {% if message %}  ❺
    <h2>{{ message }}</h2>
    {% end if %}


</body>
</html>
```

- ❶ Run through each pin in the pins dictionary.
- ❷ Print the name of the pin.
- ❸ If the pin is high, print that the device is on and link to the URL to turn it off.
- ❹ Otherwise, print that the device is off and link to the URL to turn it on.
- ❺ If a message was passed into the template, print it.

The best part about writing the code in this way is that you can very easily add as many devices that the hardware will support. Simply add the information about the device to the `pins` dictionary and add the correct action for the individual pin in each function ("ON", "OFF", or "TOGGLE". When you restart the server, the new device will appear in the status list and its control URLs will work automatically.

There's another great feature built-in: if you want to be able to flip the switch on a device with a single tap on your phone, you can create a bookmark to the address <ipaddress>/pin/toggle. That URL will check the pin's current state and switch it.

Going Further

Requests (docs.python-requests.org/en/latest)
> The home page for Requests includes very comprehensive documentation complete with easy-to-understand examples.

Flask (flask.pocoo.org)
> There's a lot more to Flask that we didn't cover. The official site outlines Flask's full feature set.

Flask Extensions (flask.pocoo.org/extensions)
> Flask extensions make it easy to add functionality to your site.

A/Writing an
SD Card Image

This book has mostly concerned itself with the Raspberry Pi OS , how to install it, and how to use it. However, as you may remember from Chapter 3, there are lots of other distributions and operating systems that can run ont he Pi.

With any distribution, it's normally just a matter of downloading the image file and then copying that file to an SD card. This appendix seeks to do two things: simplify the process of writing a disk image file to an SD card, and also to illustrate the process the other way around—to create a disk image file from an SD card. This is extremely helpful if you want to save a working SD card image, for example, so if you screw up your working Pi you can reinstall your working backup. It's also useful if you want to create an army of Pis, all running the exact same processes.

Here's how you do these things, in Windows, Mac OS, and Linux.

Writing an SD Card from OS X

There are two ways to write an image to an external device with a Mac using the terminal, or using an external application. First, let's look at the terminal method.

1. Open your Terminal application (it's in the Utilities folder inside your Applications folder.)

2. With the SD card *not* inserted, type **diskutil list** into the terminal. You'll see a list of all disk partitions currently mounted, including the main hard drive and any virtual disks you may have running (Figure A-1). Listed here, you see /dev/disk0 and /dev/disk1. The numbers below each disk correspond to disk partitions and are referred to with "s" (seen in the far right column). For example, /dev/disk1s2 is the 2nd partition of /dev/disk1.

3. Now insert the SD card and run the command again. You'll see a new entry corresponding to the card (figure A-2). Here, it's a 32GB card and is labeled /dev/disk2.

4. Unmount all partitions on the SD card by typing `sudo diskutil unmount /dev/disk2s1` (substituting whatever your particular disk and partition are labeled, of course.) If you're using a brand new SD card, it'll likely only have one partition. If it has more than one partition, make sure you unmount them all (/dev/disk2s2, /dev/disk2s3,etc.)

5. In your terminal, navigate to the location of the disk image you want to write to the card. It should have a file ending of .img; if it ends in .zip, you'll need to unzip the file first. Then write the image to the card using the following command: `sudo dd bs=4M if=filename.img of=/dev/rdisk2;sync`. This uses the Unix dd utility to write to the card, using 4MB block sizes. Note: substitute your disk label for rdisk2, make sure you put the "r" in front, and MAKE SURE YOU'RE TARGETING THE CORRECT DISK IN YOUR of=FLAG. The dd command is nicknamed "disk destroyer" because if you get the input and output file names wrong, it's very easy to write over your hard drive. Double-check everything before you hit Return. If you get the error message "dd:bs:illegalnumeric value", use `bs=1m` instead, as it depends on which version of dd you have installed on your Mac.

```
● ● ●                          wolf (-bash)                              ⌥⌘1
Galadriel:~ wolframdonat$ diskutil list
/dev/disk0 (internal, physical):
   #:                       TYPE NAME                    SIZE       IDENTIFIER
   0:       GUID_partition_scheme                       *1.0 TB     disk0
   1:                        EFI EFI                     209.7 MB   disk0s1
   2:              Apple_APFS Container disk1            1000.0 GB   disk0s2

/dev/disk1 (synthesized):
   #:                       TYPE NAME                    SIZE       IDENTIFIER
   0:       APFS Container Scheme -                     +1000.0 GB  disk1
                                 Physical Store disk0s2
   1:              APFS Volume Macintosh HD              683.1 GB   disk1s1
   2:              APFS Volume Preboot                   42.1 MB    disk1s2
   3:              APFS Volume Recovery                  510.4 MB   disk1s3
   4:              APFS Volume VM                        6.4 GB     disk1s4

Galadriel:~ wolframdonat$ █
```

Figure A-1. *diskutil list on Mac OS*

```
● ● ●                          wolf (-bash)                              ⌥⌘1
Galadriel:~ wolframdonat$ diskutil list
/dev/disk0 (internal, physical):
   #:                       TYPE NAME                    SIZE       IDENTIFIER
   0:       GUID_partition_scheme                       *1.0 TB     disk0
   1:                        EFI EFI                     209.7 MB   disk0s1
   2:              Apple_APFS Container disk1            1000.0 GB   disk0s2

/dev/disk1 (synthesized):
   #:                       TYPE NAME                    SIZE       IDENTIFIER
   0:       APFS Container Scheme -                     +1000.0 GB  disk1
                                 Physical Store disk0s2
   1:              APFS Volume Macintosh HD              683.1 GB   disk1s1
   2:              APFS Volume Preboot                   42.1 MB    disk1s2
   3:              APFS Volume Recovery                  510.4 MB   disk1s3
   4:              APFS Volume VM                        6.4 GB     disk1s4

/dev/disk2 (internal, physical):
   #:                       TYPE NAME                    SIZE       IDENTIFIER
   0:       FDisk_partition_scheme                      *31.9 GB    disk2
   1:           Windows_FAT_32 NO NAME                   31.9 GB    disk2s1

Galadriel:~ wolframdonat$ █
```

Figure A-2. *diskutil list on Mac OS*

When it's finished, you should have a working SD card. If you want to make a copy of a working Pi SD card, it's a similar process.

1. Run `diskutillist` before and after inserting the SD card to determine its label.

2. If you're using a working Raspberry Pi SD card, it'll have two partitions on it. Use `sudo diskutil unmount` to unmount both partitions.

3. Now use this command to write the disk image to a file: `sudoddbs=4Mif=/dev/rdisk2of=filename.img;sync`. Again, use your particular disk label, and note that no partitions are listed. This ensures you're copying the entire disk.

Now you have a working copy that you can then write to another card. It's a handy way of keeping a backup of a known-working system.

If you want to use a third-party utility to write to your SD card, there are several to choose from. The Raspberry Pi Foundation offers one as a freedownload—the Raspberry Pi Imager, which is available on their main software page (www.raspberrypi.org/software). I have also had very good luck with balenaEtcher (www.balena.io/etcher/). Simply download it and install it and follow the onscreen instructions to create the SD card. balenaEtcher will not let you create a disk image file, but it will let you clone an SD card directly to another card, so that's almost as helpful. Use the terminal trick above if you want to create an image file to use later.

Writing an SD Card in Linux

Although you can use third-party utilities in Linux to write image-files as well, I just use the command line, as it's very simple.

1. With the card not inserted, in your terminal, run `lsblk` (figure A-3).

2. Insert the card and run `lsblk` again (figure A-4). Note the new disk listed. In this case, the new disk is /dev/sdc and is showing as 7.5GB. Yours maybe /dev/sdb, /dev/sdc, or something else. The individual partitions of the disk are the numbers following it: sdb1, sdb2, and so on.

3. Unmount all partitions on the disk using `sudo umount /dev/sdc1`, `sudo umount /dev/sdc2`, and so on until all partitions are unmounted.

4. Now navigate, in your terminal, to the location of your disk image file. Copy it to the SD card with the following command: `sudo dd bs=4M if=filename.img of=/dev/sdc`

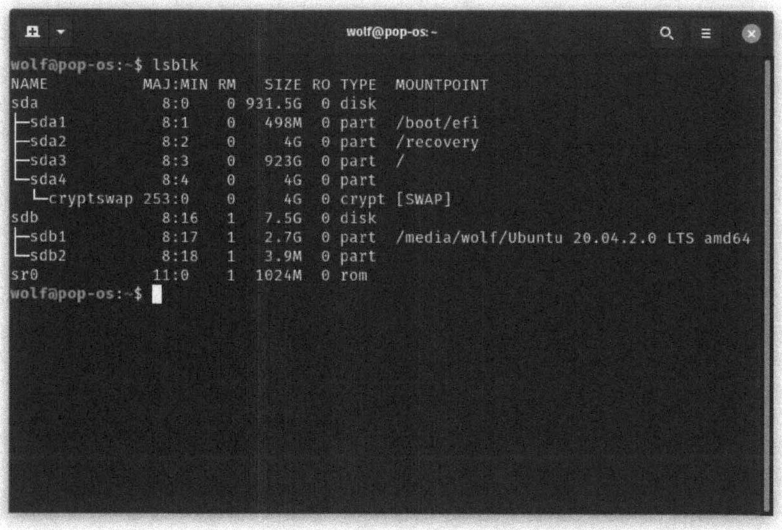

```
[⬛ ▼]                      wolf@pop-os: ~                    Q  ≡  ⊗

wolf@pop-os:~$ lsblk
NAME          MAJ:MIN RM   SIZE RO TYPE  MOUNTPOINT
sda             8:0    0 931.5G  0 disk
├─sda1          8:1    0   498M  0 part  /boot/efi
├─sda2          8:2    0     4G  0 part  /recovery
├─sda3          8:3    0   923G  0 part  /
└─sda4          8:4    0     4G  0 part
  └─cryptswap 253:0    0     4G  0 crypt [SWAP]
sdb             8:16   1   7.5G  0 disk
├─sdb1          8:17   1   2.7G  0 part  /media/wolf/Ubuntu 20.04.2.0 LTS amd64
└─sdb2          8:18   1   3.9M  0 part
sr0            11:0    1  1024M  0 rom
wolf@pop-os:~$ ▊
```

Figure A-3. *lsblk on Linux*

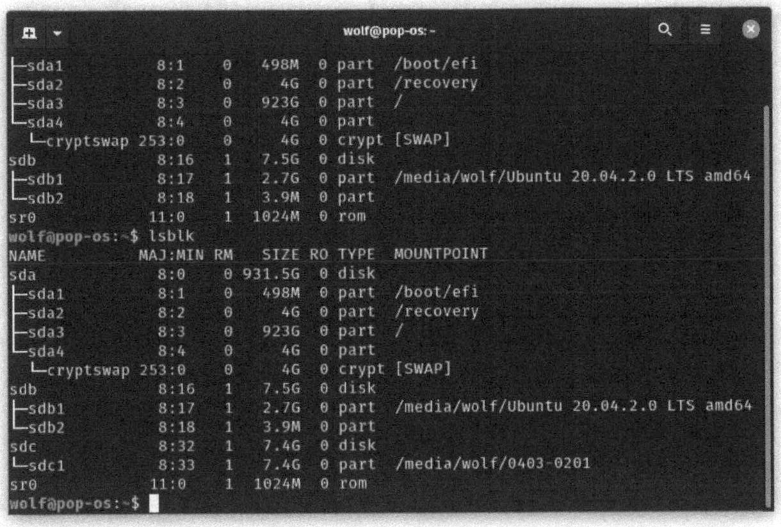

Figure A-4. *lsblk on Linux*

When the operation finishes, you'll have a working SD card. If you want to go the other way and create an image file for use later, it's just as simple.

1. Run `lsblk` before and after inserting the card to determine the correct disk label.

2. Unmount all partitions on the card. A working Raspberry Pi SD card will have two partitions; make sure you unmount them both.

3. Navigate in your terminal to the location where you would like to store the image file.

4. Run the following command to create the file, substituting your disk label: `sudo dd bs=4M if=/dev/sde of=filename.img conv=fsync status=progress`. Note that you do not specify partitions of the SD card, which ensures that you're copying the entire card, not just a single partition.

When the process finishes, you'll have an image file you can then burn to another card.

Writing an SD Card from Windows

Writing an SD card in Windows is pretty simple, as you have your choice of various third-party applications to use. The Raspberry Pi Imager (www.raspberrypi.org/software) is available for Windows, as is balenaEtcher (balena.io/etcher). I've also had good luck with Win32 Disk Imager (sourceforge.net/projects/win32diskimager/files/latest/download). Choose your favorite, install it, and follow the onscreen instructions to write to your card. I believe none of these tools will allow you to create an image file from an SD card, so you may want to switch to a Mac or Linux box if that's your goal.

B/The Raspberry Pi Pico

In this appendix, we'll take a look at the Raspberry Pi Pico. The Pico is the newest entry in the Raspberry Pi ecosystem and was released in February 2021. It's a small microcontroller, similar to an Arduino, that utilizes the RP2040 chip developed by the Raspberry Pi organization. It retails in the United States for $4. The RP2040 is a dual-core ARM Cortex-M0+ that runs at up to 133MHz and is programmable using both C and MicroPython.

The Pico Itself

Figure B-1. *The Raspberry Pi Pico*

Why, Pi?

One wonders why the Raspberry Pi Foundation would go through all the trouble of creating and designing its own micro-processing chip. We're sure there are multiple answers to that question, chief among which is the desire to simply have their own reliable source of microprocessors. But it is curious as to why Raspberry Pi would want to step into an already saturated microcontroller market with what is, essentially, a high-end Arduino.

The Pico board itself houses the RP2040 and adds 2MB of flash memory, integrated USB connectivity, and careful power management. (The board can run on as little as 1.8V, meaning it can be powered by a lithium-ion battery, or even a couple of AA batteries). There are 26 GPIO pins rated at 3.3V. With careful programming, you can have two channels of I^2C, two channels of SPI, and 16 pulse-width-modulation (PWM) pins. (The PWM pins on the Pico are interesting in that they can also measure *incoming* PWM signals, returning information on the frequency and the duty cycle of the pulses they receive.)

One interesting note about the Pico is that there is an internal ROM space that contains the chip's bootloader. This means that it should be almost impossible for you to brick the Pico, rendering it inoperable due to your bad, buggy programming. Then again, at $4 each, these boards are practically disposable, if you don't care about the environment.

The Raspberry Pi Foundation's website, raspberrypi.org, is loaded with technical documentation for the Pico and the RP2040, containing comprehensive datasheets for both devices, as well as CAD files for recreating compatible boards. It even includes the complete contents of the onboard boot ROM! There's also serious amounts of information on how to set up the entire C/C++ compilation tool chain for the Pico. All of that is beyond the scope of this book: we'll spend this chapter getting you introduced to MicroPython on the Pico.

MicroPython

MicroPython is, as the name suggests, a version of Python that is optimized to run on microcontrollers. It was created by Australian programmer Damien George in 2014, after a successful Kickstarter campaign, as a way to easily program microcontrollers and other memory-and power-constrained devices. Since the original Python was developed for full-fledged computers with adequate amounts of memory and an operating system, MicroPython has to walk a fine line between features and operability.

Step 1 on the Pico will be using MicroPython inside an interactive environment called a REPL (often pronounced with a vowel sound halfway between "repple" and "ripple"), which stands for "Read-Evaluate-Print-Loop". In a REPL, the computer first pauses to *read* input from the user, in this case, the typed message "Hello, Pico":

```
Welcome to minicom 2.7.1

OPTIONS: I18n
Compiled on Aug 13 2017, 15:25:34.
Port /dev/ttyACM0, 15:49:54

Press CTRL-A Z for help on special keys

>>> print('Hello, Pico')
```

Figure B-2. *REPL step 1: Read*

Step 2 comes when the user presses return or enter. The environment *evaluates* the input. This is where the action happens: if the command were to light up an LED or open a network connection, that would take place during the evaluation stage.

Step 3 is *print.* This is where the MicroPython environment returns the result of the evaluation stage.

```
Welcome to minicom 2.7.1

OPTIONS: I18n
Compiled on Aug 13 2017, 15:25:34.
Port /dev/ttyACM0, 15:49:54

Press CTRL-A Z for help on special keys

>>> print('Hello, Pico')
Hello, Pico
>>>
```

Figure B-3. *REPL step 3: Print*

After the statement is evaluated and printed, the REPL environment *loops* back to the first step, and waits to read a new statement
from the user. In MicroPython, this is shown by a new >>> prompt.

Figure B-4. *REPL step 4: Loop*

Like any good Python variant, the MicroPython REPL will auto-
indent your code as you type, and even offers auto-complete by
pressing the TAB key as you type your command. As with regular
Python, hitting the Ctrl-C key combination is supposed to interrupt
a running program. If that doesn't work, or if you need to completely
reset the REPL, the Ctrl-D combination will restart the MicroPython
interpreter without disconnecting the link between the Pico and
your computer. Usually.

Some Differences between MicroPython and Python:

- MicroPython requires spaces between literal numbers and keywords

- Unicode name escapes are not implemented

- Error messages for methods may display unexpected argument counts

- Function objects do not have the module attribute

- User-defined attributes for functions are not supported

- Failed to load modules are still registered as loaded

If you think about it, nearly all of these differences are caused by the need to keep MicroPython as compact (and memory-kind) as possible.

Installing MicroPython on the Pico

The Raspberry Pi operating system came equipped with Python. The Pico doesn't come equipped with MicroPython, but adding it is a snap.

1. Completely unplug your Pico from all sources of power. For-most users, this simply means unplugging the USB cable from your computer. It also means disconnecting any LiPo or other batteries you may be using. For this to work, the Pico must be absolutely dead.

2. On your computer, go to micropython.org/down-load/rp2-pi-co/rp2-pico-latest.uf2 to download the latest MicroPython code for the Pico.

3. Press and hold the BOOTSEL button on the Pico.

Figure B-5. *The BOOTSEL (Boot Select) button on the Raspberry Pi Pico*

4. Power up the Pico by plugging in the USB cable attached to your computer.

5. Release the `BOOTSEL` button.

6. On your computer, you should see the Raspberry Pi Pico appear in your file manager, as a USB Mass Storage Device called `RPI-RP2`. (It may take a few moments for it to register.) If you examine it with some drive manager software, you'll see that the Pico pretends to be a 128MB FAT flashdrive.

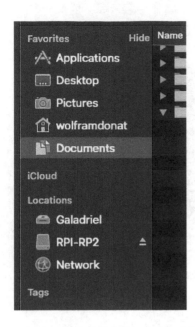

Figure B-6. *The Pico as a USB drive attached to your computer*

7. Use the file manager to drag and drop the `rp2-pico-latest.uf2` file from your computer to the Pico, just as you would move any other file from one drive to another.

8. The Pico will ingest the file (it may take a few seconds), and automatically reboot. Your Pico is now running the MicroPython REPL, and is no longer a USB drive.

9. At this point, your Pico is now a character device attached to your computer's serial port. To communicate with the Pico's REPL, you'll need a terminal emulator program.

Linux and Mac

In their documentation, the Raspberry Pi Foundation recommends using `minicom`, a text-based serial terminal emulator, but really, just about any terminal emulator like `CoolTerm` or `PuTTY` will do.

First, you'll have to determine which serial port your Pico is attached to. If you're working in Linux, you can simply run `ls/dev/tty*` and look for `/dev/ttyACM0`. (This includes if you're working on a Raspberry Pi; it's definitely possible to work with your Pico from your Pi!)

```
pi@raspberrypi:~ $ ls /dev/tty*
/dev/tty     /dev/tty19   /dev/tty3    /dev/tty40   /dev/tty51   /dev/tty62
/dev/tty0    /dev/tty2    /dev/tty30   /dev/tty41   /dev/tty52   /dev/tty63
/dev/tty1    /dev/tty20   /dev/tty31   /dev/tty42   /dev/tty53   /dev/tty7
/dev/tty10   /dev/tty21   /dev/tty32   /dev/tty43   /dev/tty54   /dev/tty8
/dev/tty11   /dev/tty22   /dev/tty33   /dev/tty44   /dev/tty55   /dev/tty9
/dev/tty12   /dev/tty23   /dev/tty34   /dev/tty45   /dev/tty56   /dev/ttyACM0
/dev/tty13   /dev/tty24   /dev/tty35   /dev/tty46   /dev/tty57   /dev/ttyAMA0
/dev/tty14   /dev/tty25   /dev/tty36   /dev/tty47   /dev/tty58   /dev/ttyprintk
/dev/tty15   /dev/tty26   /dev/tty37   /dev/tty48   /dev/tty59   /dev/ttyS0
/dev/tty16   /dev/tty27   /dev/tty38   /dev/tty49   /dev/tty6
/dev/tty17   /dev/tty28   /dev/tty39   /dev/tty5    /dev/tty60
/dev/tty18   /dev/tty29   /dev/tty4    /dev/tty50   /dev/tty61
pi@raspberrypi:~ $ 
```

Figure B-7. *Listing/dev/tty* from a Pi*

Microsoft Windows

In Windows, you'll have to open your device manager:

Figure B-8. *Device manager*

And then under the `View` menu, sort the devices by container.

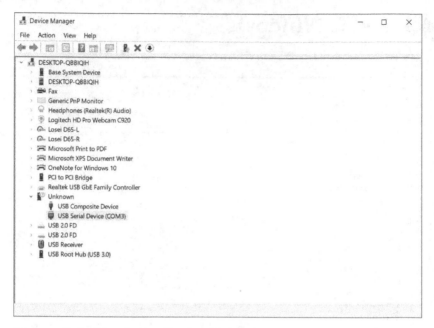

Figure B-9. *Pico in the Device Manager*

Look for the device that's a "Board in FS Mode". (In some instances, you may see a message on plugging in the Pico that "Windows is setting up a Board in FS Mode," but then the board will be populated in Device Manager as a USB Serial Device under "Unknown" in the main window. Note the specific COM port it's using, COM3 in this case.

Still in Device Manager, click on the entry for COM3 and choose the Port Settings tab.

Figure B-10. *Settings for COM3 connection*

Note down the communications parameters: 9600 bits per second, 8 data bits, no parity, 1 stop bit, no flow control.

Then open your terminal emulator. We're using PuTTY for Windows. Enter the communications parameters into the settings, making sure you're specifying a Serial connection rather than an SSH one, and you should be connected to the Pico's MicroPython REPL.

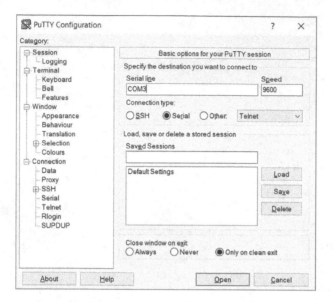

Figure B-11. *Entering settings in PuTTY*

If you're using Linux (on your Pi, for example), you don't need any of the communications parameters, as the Pico is really attached to a virtual serial port. If you're using minicom, just enter `minicom-o-D/dev/ttyACM0` and press Enter once the minicom window opens to get the initial `>>>` REPL prompt. (Minicom can be installed with a simple `sudo apt-get install minicom`.)

Using MicroPython on the Pico

So now your computer is connected directly to the Pico's REPL via your terminal emulator program. You should see something like this:

Figure B-12. *Minicom connection to the Pico onthe Pi*

It looks almost exactly like the description of the REPL a few pages ago, because it is. You can stick to tradition and type print ("Hello World") as your first MicroPython program if you want to, but there are more exciting things in store.

Blinking an LED on the Pico

```
from machine import Pin
led = Pin(25, Pin.OUT)
led.value(1)
```

What happened? One of the LEDs on the Pico just lit up! To shut it off, type into the REPL:

```
led.value(0)
```

It's generally accepted that the ability to control a microcontroller pin in this way is the foundation of all physical computing. Now that you've got that pin working, the world is yours!

Digression

Did you know that the Chrome browser keeps track of the devices plugged into your computer? And did you know that you can get Chrome to tell you which devices are plugged into your computer? Just type: chrome://device-log/?refresh=30 into the address bar to see what Chrome knows about your computer. Kind of cool, and kind of creepy.

C/Another Raspberry Pi?!

Figure C-1: *The Raspberry Pi Zero W2*

As this book goes to press, the Pi Foundation has just announced another entry in their lineup of Pi models: a souped-up model Zero W, known as the Zero W2.

So what's special or interesting about the Zero W2? As a Pi Zero model, it's about the size of a stick of gum — less than half the size of a normal Pi. It's also about a third of the price of the new Pi 4. At the $15 price point, it's ideal for throwing a prototype together or putting it into a set-and-forget thingamajig in which you wouldn't want to use a more expensive, full-size Pi. It's like the Zero W in that it has onboard Wi-Fi, and, get this, the chip it uses, the RP2041, contains the same processor as the Raspberry Pi 3.

In case you *didn't* get that, let us reiterate: this is basically a Raspberry Pi model 3 (which is still a *perfectly good*, capable Pi) but smaller, and it costs only 15 bucks. It's been under-clocked a bit to 1GHz instead of the Pi 3's 1.2GHz, but it's the same processor. (We have a feeling that some of us are going to be experimenting to see if we can overclock it and recover those lost clock cycles.)

There are a few other products in the Pi Foundation's lineup that come close to the Zero W2 in terms of size and power. The one that comes to mind first is the Compute Module 3. The Compute Modules tend to be forgotten by many makers when it comes to building things because they're not quite as user-friendly. They don't have connectors for HDMI and USB, or an SD card slot, as they're designed to plug into another board with a single edge connector. Plus, when you add the cost of the development board to which you're connecting, the price increases quite a bit. But power- and size-wise, the Compute Module 3 is pretty close to the Zero W2.

So what's the big deal about the Zero W2? After all, there's already a Zero and a Zero W. First, if you've looked around recently, you may have noticed it can be difficult to find a Zero. Maybe it's because there just aren't many around, or it may be that they're very popular, but searching for a Pi Zero on sites such as Amazon or Pimoroni will often bring up the Zero W but not the Zero. The Zero has one major pro and one major con. The pro — the price. It's hard to beat a Pi for $5. The con — no Wi-Fi capability built in. If you want to connect the Zero to a network, you have to use a USB-to-Ethernet adapter, which either takes one of the precious ports on the board or requires a USB hub, and both of those take up space that you may not be ready to lose in the guts of your prototype.

The next step up, obviously, is the Zero W. We've always assumed that users just prefer the Zero W because of its inbuilt Wi-Fi and Bluetooth connectivity; many makers want to build connected devices, and a device that is already set up to connect to Wi-Fi just makes sense. The Zero W is only $10, but it has one drawback, and that is supply. As of this writing, all suppliers selling the Zero W limit customers to one, and this has not changed in the more than four years since it was first released. (I was recently experimenting with a sort of Zero

W Beowulf cluster and was forced to make lots of small individual purchases to obtain my cluster of Ws.)

Now, makers have the choice of using the Zero W2, and it's almost a no-brainer. (Of course, we'll have to see what the supply/demand issues work out to be, since it's bound to be a popular board.) In testing, the Zero W2 is light years faster than the Zero. This is most likely due to the Zero W2's multicore processor, which can be taken full advantage of using multi-threaded code.

It's still not going to be anybody's desktop computer, since it's so lightweight and will likely have trouble with any Javascript-heavy website that it visits. So this is most likely going to be used in projects and designs and inventions. But that's exactly what the Pi was originally supposed to be used for. It's kind of nice, then, to see this small, powerful board that's able to really boost the horsepower in small, embedded systems. It's also encouraging that the onboard chip, the RP2041, is the Pi Foundation's second microcontroller chip. We'll be curious to see what they'll come up with next, since they seem to be doing quite well making their own silicon.

Oh, one last thing: we can't guarantee results, but most if not all of the code in this book should run on the Zero W2, with the possible exception of the OpenCV chapter. Again, we're revisiting the Pi's roots as a low-cost teaching computer, and you can't really beat $15.

Have fun programming!

Index

Symbols

(hash) tags and cron, 105
-- help option, 39
- a switch, 37
. (dot), 36
.. (dots), 36
802.11 Wi-Fi USB dongles, 10
| (pipe) operator, 39

A

absolute paths, 36
Adafruit Industries
    ADS1115 breakout board for, xviii,
    128
    online store, 92
ADS1115 (Texas Instruments),
    128-133
Advanced Linux Sound Architecture
    (ALSA), 48
AlaMode shield (WyoLum), 88
    alsa mixer program, 48
Amiga, ix
analog input/output, 123-138
    for Arduino, 86
    converting from digital, 124
    converting to digital, 128-134
    variable resistors, 134
analog-to-digital converter (ADC),
    128-134
Apache webserver, 177
API (application programming inter-
    face), 169,176
apt-get command-line utility, 48
ArchLinux, 52
Arduino, 77-88, 198
    communicating with, 80-84
    default font, changing, 80
    environment location, 78

Firmata and, 86
installing, 79
powering, 80
PWM on, 127
Python in, 62
on serial port, finding, 81
serial protocols, 88
tutorials webpage, 78
user experience, improving, 82
ARM, 197
associative arrays (Python), 72
Atom feeds, grabbing, 71
Atrix lapdock, 11
audio out (Raspberry Pi), 4
    forcing output to, 48
    omxplayer, 146
sending sound to, 120
autocomplete, 33
avahi-daemon, 178

B

Banzi, Massimo, 78
bare-metal computer hacking, xiv
Barrett, Daniel J, 50
bash terminal shell, 33
BASIC, programming language, 55
BBC Micro, x
BBC News, x
Behringer's U-Control devices, 11
Berdahl, Edgar, 54
BitTorrent, 16
blit function, 162-164
Bonjour Services, 178
Briggs, Jason R., 75
Broadcom chipset, 14
BUB I board (Modern Device), 25
buttons (physical)
    and breadboards, 98
    reading, 98-102
    reading in Python, 114-116
    updating websites with, 181

C

camera module (Raspberry Pi), xvii,
    7-8, 11, 139-166
Camera Serial Interface (CSI) con-

M

Maker Shed online store, 92
Making Things Talk, 2E (Igoe), 88
man command (Linux), 39
microcontroller, 197
microcontrollers, Raspberry Pi vs., xiii
MIDI protocol, 88
Miro BitTorrent client, 16
MLDonkey BitTorrent client,16
Model A (Raspberry Pi)
  audio/video outs, 4
  keyboard/mouse, plugging in, 16
  USB ports, 3-5
Model B (Raspberry Pi)
  audio/videoouts, 4
  USB ports, 3-5
Model B+ (Raspberry Pi)
  audio/video outs, 4
  GPIO pins on, 5
  HATs for, 12
  USB ports, 3-5
Modern Device, 25
modules (Python), 65-69
  importing,67
  user-defined, 67
Mouser online store, 92
multimeter, 137
music distributions, 54
Mustacheinator project (Practical Computer Vision with SimpleCV), 160
mv command (Linux), 38

N

network connectivity, 22, 46
Network Time Protocol (NTP) server, 47
New Hackers Dictionary (website), 47
Node.js protocol, 88
NOOBS installer, 15, 32, 52

O

objects (Python), 65-68
OctoPi distribution, 57

Oliver, Anthony, 160
omxplayer, 146
Oostendorp, Nathan, 160
Open Embedded Linux Entertainment Center distribution, 53
OpenELEC, xiv
OpenELEC distribution, 53
Openwrt distribution, 57
operating system development, xiv
operating systems
  booting, 16
  distributions, 14
  installing from SD cards, 15
OSX
  sharing Wi-Fi with, 22-23
  SSH utility, 24
  writing SD cards in, 190
OSMC distribution, 52
OSMC open source media player, xiv
overclocking processors, 20
overscan option, 19

P

package managers, 47
passwd command (Linux), 45
password, setting, 19, 45
pegging the processor, 120
peripherals, 8-12
permissions
  in Linux, 48
  for RPi.GPIO, 109
  for serial port, 81
Photobooth project, 160-165
photocells, 134
Pi Cobbler Breakout Kit (Adafruit), 92
pi user account, 43
Pibow case, 13
Pidora, 52
PiMAME distribution, 55
Pimoroni, 13
ping command (Linux), 46
Pip, installing, 71
pipes, in Linux, 39
PiPlay distribution, 55
potentiometers, 129, 134
power supply, 8

X

About the Authors

Matt Richardson is an Executive Director for the Raspberry Pi Foundation and is responsible for their non-profit work within North America. He's a graduate of New York University's Interactive Telecommunications Program. Highlights from his work include the Descriptive Camera (a camera that outputs a text description instead of a photo) and The Enough Already (a DIY celebrity-silencing device). Matt's work has been featured at The Nevada Museum of Art, The Rome International Photography Festival, and Milan Design Week, and has garnered attention from The New York Times, Wired, and New York Magazine.

Shawn Wallace lives in Providence, RI, and builds creative coding tools for young people at Unruly Studios. He is the inventor of Fluxly, Cryptozoologic, and the Fluxamasynth. Previously he helped start the Providence FabLab, wrote and edited books for O'Reilly and Maker Media, and designed electronics for Modern Device.

Wolfram Donat is an engineer, maker, and author who has been building things with the Raspberry Pi since he got his first Model 1A+ delivered, oh-so-many moons ago. He's currently the Software Architect at Arc Machines, Inc., where he uses the Pi (among other things) to build and control intelligent welding machines. This marks his fourth foray into books about/utilizing the Raspberry Pi.

Colophon

The cover and body font is Benton Sans, the heading font is Serifa, and the code font is The Sans Mono Condensed.

CPSIA information can be obtained
at www.ICGtesting.com
Printed in the USA
JSHW021735021121
20081JS00001B/1

9 781680 456998